T0205612

Cognitive Biases in Visualizations

Geoffrey Ellis
Editor

Cognitive Biases
in Visualizations

 Springer

Editor
Geoffrey Ellis
Data Analysis and Visualisation Group,
    Department of Computer and Information
    Science
University of Konstanz
Konstanz, Germany

ISBN 978-3-030-07104-2   ISBN 978-3-319-95831-6   (eBook)
https://doi.org/10.1007/978-3-319-95831-6

This Springer imprint is published by the registered company Springer Nature Switzerland AG
The registered company address is: Gewerbestrasse 11, 6330 Cham, Switzerland

Preface

… and their judgement was based more upon blind wishing than upon any sound prevision; for it is a habit of mankind to entrust to careless hope what they long for and to use sovereign reason to thrust aside what they do not fancy. Thucydides 420 BC

Little did the Greek Historian Thucydides know that he was describing, what is essentially confirmation bias, almost 2500 years ago. People were making poor decisions then and still are … and at least some of these are due to cognitive biases and are not necessarily their fault. It's only since the 1970s that the term cognitive bias has been used to describe "errors in judgment" or "irrational choices". Hundreds of cognitive biases have been described in terms of people's decision-making behavior in particular situations, however there has been limited progress in mitigating the impact of cognitive biases and hence improving judgements. Some five years ago, whilst working on a project concerned with cognitive biases, I speculated whether people's decision-making, whilst using visualisation applications, are subject to cognitive biases, and if so, can we adapt such applications to lessen their impact and hence improve decisions. This lead to the 1st DECISIVe workshop at IEEE 2014 in Paris, *Dealing with Cognitive Biases in Visualisations*, with the aim of providing a forum for researchers and practitioners, from a wide range of disciplines, to raise and discuss pertinent issues concerning cognitive biases in visualizations. This raised awareness in the subject area and kick-started research. The 2nd DECISIVe workshop took place at IEEE VIS 2017, in Phoenix, USA with the aim to highlight ways in which cognitive biases have a detrimental impact on users decision making when using visualisation and analytics tools, and to explore practical ways to "measure" the occurrence of cognitive biases and develop ways of reducing their potentially harmful effects. Accepted papers from this workshop form the basis of this book.

DECISIVe 2017 Workshop

Original submissions were typically 5 pages in IEEE format and were carefully reviewed by three program committee members. Seven submissions were accepted with minor corrections, whilst another seven were subject to major revisions, which were then reviewed again before final acceptance.

Eight of the workshop papers have been extended (between 50 and 100%) by their authors whilst another three have minor changes. Two chapters have been added—an introduction to cognitive biases, providing background to the subject area, and an invited contribution from Donald Kretz, who has 30 years' experience in cognitive science research, decision making and intelligence analysis.

Organizers

Geoffrey Ellis, Data Analysis and Visualisation Group, Department of Computer and Information Science, University of Konstanz, Germany
Evanthia Dimara, AVIZ team, INRIA, Paris, France
Donald Kretz, Security Engineering and Applied Sciences, Applied Research Associates, Albuquerque, USA
Alex Endert, Visual Analytics Lab, School of Interactive Computing, Georgia Institute of Technology, Atlanta, Georgia, USA

Program Committee Members

Simon Attfield, Department of Computer Science, Middlesex University, UK
Gilles Bailly, Institute for Intelligent Systems and Robotics, Sorbonne Université, Paris, France
Emmanuelle Beauxis-Aussalet, Centrum Wiskunde & Informatica, The Netherlands Organisation for Scientific Research, Amsterdam
Roger Beecham, Geographic Data Science, Leeds University, UK
Michael Correll, Tableau Research, Seattle, Washington, USA
Pierre Dragicevic, AVIZ, INRIA, Paris, France
William Elm, Resilient Cognitive Solutions, USA
BC Kwon, IBM T.J. Watson Research Center, New York, USA
Gaelle Lortal, Thales, Paris, France
Alexander Nussbaumer, Institute of Interactive Systems and Data Science, Graz University of Technology, Austria
Margit Pohl, Institute of Design and Assessment of Technology, Vienna University of Technology, Vienna, Austria
Dominik Sacha, Data Analysis and Visualisation Group, Department of Computer and Information Science, University of Konstanz, Germany
David Peebles, Applied Psychology, University of Huddersfield, UK
Connor Upton, Fjord, Dublin, Ireland

I would like to thank the authors of the chapters for their valuable contributions, the assistance from the DECISIVe organizing committee and the commitment from the program committee during the reviewing process. Thanks also to Helen Desmond from Springer for her help and support in preparing this book.

I would also like to acknowledge the support given by the EU project VALCRI under grant number FP7-SEC-2013-608142.

Konstanz, Germany   Geoffrey Ellis

Contents

Part III   Mitigation Strategies

Contributors

Dietrich Albert Institute of Interactive Systems and Data Science, Graz University of Technology, Graz, Austria

Rahul C. Basole Georgia Institute of Technology, Atlanta, GA, USA

Michael A. Bedek Institute of Interactive Systems and Data Science, Graz University of Technology, Graz, Austria

Leslie M. Blaha Pacific Northwest National Laboratory, Richland, WA, USA

André Calero Valdez Human-Computer Interaction Center, RWTH Aachen University, Aachen, Germany

Min Chen Oxford e-Research Centre at the University of Oxford, Oxford, UK

Kristin Cook Pacific Northwest National Laboratory, Richland, WA, USA

Joseph A. Cottam Pacific Northwest National Laboratory, Richland, WA, USA

Aritra Dasgupta New Jersey Institute of Technology, New Jersey, USA

Geoffrey Ellis Data Analysis and Visualization Group, University of Konstanz, Konstanz, Germany

Alex Endert Georgia Institute of Technology, Atlanta, GA, USA

Lane Harrison Worcester Polytechnic Institute, Worcester, MA, USA

Luca Huszar Institute of Interactive Systems and Data Science, Graz University of Technology, Graz, Austria

Daniel A. Keim Data Analysis and Visualization Group at the University of Konstanz, Konstanz, Germany

Donald R. Kretz Security Engineering and Applied Sciences, Applied Research Associates, Inc, Frisco, TX, USA; School of Behavioral and Brain Sciences, University of Texas at Dallas, Richardson, TX, USA

Po-Ming Law Georgia Institute of Technology, Atlanta, GA, USA

Hamid Mansoor Worcester Polytechnic Institute, Worcester, MA, USA

Ronald Metoyer University of Notre Dame, Notre Dame, IN, USA

Alexander Nussbaumer Institute of Interactive Systems and Data Science, Graz University of Technology, Graz, Austria

Paul Parsons Purdue University, West Lafayette, IN, USA

Celeste Lyn Paul U.S. Department of Defense, Washington, D.C., USA

Margit Pohl Institute of Design and Assessment of Technology, Vienna University of Technology, Vienna, Austria

Michael Sedlmair Jacobs University Bremen, Bremen, Germany

Dirk Streeb Data Analysis and Visualization Group at the University of Konstanz, Konstanz, Germany

Poorna Talkad Sukumar University of Notre Dame, Notre Dame, IN, USA

Emily Wall Georgia Institute of Technology, Atlanta, GA, USA

Martina Ziefle Human-Computer Interaction Center, RWTH Aachen University, Aachen, Germany

Chapter 1
So, What Are Cognitive Biases?

Geoffrey Ellis

1.1 Introduction

Decisions, decisions, decisions, we make them all the time, probably thousands each day. Most are part of daily living, such as moving about our environment, others need more thought, but are not particularly critical, such as what coffee to buy. However, some decisions are important, even with life implications, from deciding if it's safe to cross the road, to a doctor deciding what cancer treatment to suggest for a patient. We might imagine that all these decisions, whether trivial or not, are based on sound reasoning using our senses and our experience stored in our memory. However, it is generally agreed that the majority of decisions are made unconsciously using heuristics - strategies that use only a fraction of the available information. This makes sense in evolutionary terms [32], as to survive approaching danger, for instance, decisions had to be made rapidly. Humans do not have the time or brain processing power to do much else than use heuristics, and are, in fact, inherently lazy in order to conserve precious energy resources [22]. Fortunately, most of the time the result of using the very fast and automatic heuristic strategies are "good enough", however in certain situations they are not good enough, leading to poor judgments. It is these "errors in judgment" or "irrational behavior" that are commonly referred to as *cognitive biases*.

During this decade, interest in cognitive biases has increased markedly, with several large research projects [38, 57] starting in 2012, as well as a few mentions in popular on-line publications [8] and even in the press. In addition to an increase in scholarly articles,[1] the biggest change has been in media interest, especially in the business world. A recent Google search for "cognitive bias" presents many business orientated items which are either aimed at selling (e.g. *Cognitive Biases : How to*

---

[1] A Google scholar search for "cognitive bias" reports 3000 in 2012 and 5480 in 2017.

G. Ellis (✉)

Data Analysis and Visualization Group, University of Konstanz, Konstanz, Germany

e-mail: ellis@dbvis.inf.uni-konstanz.de

© Springer Nature Switzerland AG 2018

G. Ellis (ed.), *Cognitive Biases in Visualizations*,

https://doi.org/10.1007/978-3-319-95831-6_1

Use Them to Sell More) or as a warning (e.g. *Hidden Cognitive Biases That Cost You Big Money*). Other search results are generally pessimistic regarding cognitive biases such as *The 17 Worst Cognitive Biases Ruining Your Life*!

More recently, *implicit* or *unconscious bias* has been in the media, in the context of equality and anti-discrimination. This is often the result of stereotyping which is influenced by our background, culture and experience. In this sense "unconscious" means that humans make this judgment without realizing it, as with heuristic processing. And, if we think that cognitive biases only affect humans, then there are studies on rats [6], sheep [69], bees [62], chicken [72] and many other animals which use cognitive bias as an indicator of animal emotion [59]. However, these uses of the term "cognitive bias" differ from the more traditional one which we are discussing in this book.

Before considering cognitive biases (in humans) in the context of visualization and visual analytics tools, the next sections, provide some examples of common cognitive biases and a brief history of their 'discovery' and subsequent research.

1.1.1 Examples

A recent classification of cognitive biases, the Cognitive Bias Codex by Benson [47] lists 187 biases.[2] These have been added to since the 1970s and the trend seems to be continuing, although sometimes just a bias by another name. There are, of course, similarities which various classification schemes over the years have attempted to tease out [3, 4, 9, 34, 37, 58, 66, 70] although most of the work has been in the area of decision support. In Chap. 2, Calero Valdez et al. propose a framework, specifically for the study of cognitive biases in visualization, and contrast this with the aforementioned Cognitive Bias Codex.

For those readers, not familiar with cognitive biases, here are four examples of common biases:

Familiarity/availability bias is where people tend to estimate the likelihood of something to happen by how easy it is to recall similar events. For instance, people will generally think that travel by airplane is significantly more dangerous in the aftermath of a plane crash being reported in the media (see Chap. 6).

Confirmation bias is where people tend to search for confirming rather than for disconfirming evidence with regard to their own previous assumptions. For example, if you think that eating chocolate makes you loose weight then a Google search "loose weight by eating chocolate" will confirm this if you ignore article to the contrary (see Chap. 5).

Representational bias in visualization involves constraints and salience. For example, a matrix representation is not good at showing network data (a constraint) but can highlight missing relationships in its table view (salience) (see Chap. 10).

---

[2]The author's own survey collected 288 distinct biases.

Overconfidence bias is where people tend to assess the accuracy of their answers or performance as greater than it actually is. There are many related cognitive biases such as illusion of control and planning fallacy (see Chap. 9).

1.2 A Brief History of Cognitive Biases

Early research on decision-making was founded on the theory of rational choice, where a person carefully assess all the alternatives and if they make errors these would not be systematic. However, in the 1950s and 60s, experiments found that people are generally poor at applying even basic probability rules and often make sub-optimal judgments when measured against an 'ideal' standard derived from Bayesian analysis [19]. Even experts, such as physicians, were found to make biased judgments [48]. Simon proposed bounded rationality [63], suggesting that humans are too limited in their data processing abilities to make truly rational decisions but employ simplifying heuristics or rules to cope with the limitations.

In the early 70s, Tversky and Kahneman developed this idea with their *heuristics–biases* program, with particular attention on judgments involving uncertainty. Systematic deviations from 'normative' behavior were referred to as cognitive biases and this was backed up by a series of experiments which illustrated 15 biases [66]. Heuer [34] also promoted the idea of cognitive bias errors being due to irrationality in human judgment with his work amongst intelligence analysts. Over the years many more cognitive biases were proposed, based mostly on laboratory experiments. However in the 80s, researchers began to question the notion that people are error prone and a lively debate has ensued over the years typified by the communication between Gigerenzer [26], and Kahneman and Tversky [41]. One of the opposing arguments poses the question "Are humans really so bad at making decisions, especially where it involves uncertainty?". Gigerenzer [28] suggests that the use of heuristics can in fact make accurate judgments rather than producing cognitive biases and describes such heuristics as "fast and frugal" (see Chap. 13).

It is suggested that the success of *heuristics–biases* program is partly due to the persuasive nature of the experimental scenarios, often in the laboratory, which can easily be imagined by the reader [42]. However, many of the studies have clearly involved domain experts in the workplace. Another criticism of the heuristics–and–biases approach is the resultant long list of biases and heuristics, with no unifying concepts other than the methods used to discover them [4]. So the focus of later work has been to propose decision making mechanisms rather than just looking for departures from normative (ideal) models [40]. To this end, dual process models have been put forward, for example the two system theories of reasoning which feature System 1: involuntary/rapid/rule-based + System 2: conscious/slower/reasoning decision making [22, 65]. Kahneman's book "Thinking, Fast and Slow" [39] also adopts this dual process model and gives a very readable account of heuristic and biases.

Other developments include the *Swiss Army Knife* approach [29] that proposes that there are discrete modules in the brain performing specific functions, and deviations occur when an inappropriate module is chosen or where no such module exists, so the next best one is used. Formalizing heuristics [27] and modeling cognitive biases [36] are other approaches to understanding what is going on in our heads when we make decisions. A useful discussion of the impact of Tversky and Kahnemans work can be found in [24]. But as research continues in this area, Norman provides a word of warning, especially in medical diagnosis, that there is bias in researching cognitive bias [51].

1.3   Impact of Biases

Not withstanding the debate amongst researchers as to the underlying cognitive processes, there is little doubt that in particular circumstances, systematic behavior patterns can lead to worse decisions. Making a poor decision when buying a car by overrating the opinion of a person you have recently met (*vividness bias*), is often not a major problem, but in other realms such as medical judgments and intelligence analysis, the implications can be damaging. For instance, a number of reports and studies have implicated cognitive biases as having played a significant role in a number of high-profile intelligence failures (see Chap. 9). Although uncertainty is a factor, a person's lack of knowledge or expertise is not the overriding consideration. Cognitive biases such as overconfidence and confirmation are often associated with poor judgments among people in senior roles, as in a realistic study where all the twelve experienced intelligence analysts were led astray by confirmation bias, leaving only the inexperienced analyst with the correct answer [5].

In addition to Chap. 9, which focuses on intelligence analysis, many of the chapters in this book describe the impact of various cognitive biases, especially in relation to interpreting visualizations or when using visualization tools. For instance, Chap. 6 details the impact of familiarity related biases, especially with experts from the physical sciences and Chap. 10 discusses potential problems with representational biases when viewing visualizations. The case study described in Chap. 12 reveals the likelihood of numerous cognitive biases which can seriously affect decision making in a college admissions process. Chapters 3 and 4 discuss the notion that various aspects of computer systems, as well as humans, can also exhibit biases.

1.4   Cognitive Biases in Visualization

Interest in cognitive bias research has grown considerably at both the cognitive science level and also in relation to the visual analytics and decision-making tools that we build. The DECISIVe workshops[3] have focused on two main issues related

---

[3]Full papers for DECISIVe 2014 are available [20].

to visualization: (i) is the interpretation of visualizations subject to cognitive biases and (ii) can we adapt visualization tools to reduce the impact of cognitive biases?

1.4.1 Interpretation of Visualizations

There is evidence from peoples' susceptibility to optical illusions that systematic errors can occur due to simplifying heuristics, such as grouping graphic items together, as set out in the Gestalt principles [1, 53, 55, 67]. It has also been demonstrated that different visual representation of common abstract forms or appearance of the visualization itself can affect the interpretation of the data [12, 16, 54, 74, 75, 77]. In relation to the comprehension of images, Fendley [23] discusses cognitive biases in detail and proposes a decision support system to mitigate a selection of biases. Ellis and Dix [21] proposed that cognitive biases can occur in the process of viewing visualizations and present examples of situations where particular cognitive biases could affect the user's decision making. Recent studies into priming and anchoring [68], the curse of knowledge [73] and the attraction effect [17] demonstrate these cognitive bias effects when interpreting visualizations, but as their authors point out, much more work needs to be done in this area.

1.4.2 Visualization Tools

In visual analytics, user interaction plays a significant role in providing insightful visual representations of data. As such, people interact with the systems to steer and modify parameters of the visualization and the underlying analytical model. While such human-in-the-loop systems have proven advantages over automated approaches, there exists the potential that the innate biases of people could propagate through the analytic tools [61]. However, if the system is able to monitor the actions of the user and their use of the data resources, then it may be possible to guide them and reduce the impact of particular cognitive biases. This requires ways to effectively detect and measure the occurrence of a range of cognitive biases in users [10, 45, 46, 71]. Work towards this is the subject of Chaps. 5, 7 and 9 in particular. Researchers point out that novel corrective actions, ideally tailored to the user, are then required.

1.5 Debiasing

Reducing the negative impact of cognitive biases is a challenge due to the inherent nature of biases and the indirect ways in which they must be observed. Early work generally focused on developing user training, typically scenario-based, in an attempt to mitigate the effect of a small number of cognitive biases. However, this approach has met with little convincing generalizable and lasting success. Research shows that even if users are made aware of a particular cognitive bias, they are often unwilling to accept that their decisions could be affected by it, which itself constitutes bias blind spot [56]. Structured analytical techniques (SATs) (as discussed in [35]), such as 'argument mapping' and Analysis of Competing Hypotheses (ACH) have been used in intelligence analysis to reduce the impact of cognitive biases. Few of these techniques have been evaluated in empirical studies, apart from ACH, which, for realistic complex problems, has proved unsatisfactory, often due to the time pressures (see Chap. 9).

There has been appreciable effort in the medical field to identify cognitive bias effects and reduce prevalent diagnostic errors [14, 15, 30] with interventions (checklists) to increase clinicians knowledge, improve clinical reasoning and decision-making skills [11] or assist clinicians with selected tools. According to Croskerry [13], progress is being made, but this is hindered by the general lack of education in critical thinking amongst clinicians.

Bias-Reducing Analytic Techniques (BRATS) are another way of investigating bias mitigation. They benefit from minimally intrusive cognitive interventions [44] based on prior work on cognitive dis-fluency [33]. While results were mixed, opportunities for further research show promise. Another method involves the application of serious games to improve critical thinking as in the MACBETH [18] and HEURISTICA [2], games developed as part of IARPA's Sirius program [7].

A common challenge across all these methods is the difficulty to shape an individual's cognitive behavior. Therefore, research is shifting toward modifying and improving the decision environment (i.e. tools, etc.). Recent works investigate how visualizations can reduce base-rate bias in probabilistic reasoning [43, 49]. Other visualization research focuses on the cognitive biases that affect judgments under uncertainty [78]: for example in finance, helping investors to overcome uncertainty aversion and diversification bias [60] or loss aversion and conservatism [76]; assisting Fantasy baseball experts to mitigate the regression bias in their predictions [50]; or countering the anchoring and adjustment bias in decision support systems [25].

Researchers further propose frameworks, integrated into visual analytic systems, that provide support for mitigating some cognitive biases through measures such as the use of appropriate visualization types, uncertainty awareness, the use of statistical information and feedback from evidence-based reasoning [52, 61]. Other approaches attempt to "externalize the thinking" of the decision-maker [45] or improve hypothesis generation [31], in this case to avoid confirmation bias.

1.6 Conclusion

Cognitive biases are still somewhat intriguing. How humans actually make decisions is still largely a mystery, but we do know that most of this goes on at an unconscious level. Indeed, neuroscience experiments suggest that human decisions for physical movement are made well before the person is consciously aware of them [64]. From a survival of the species point of view, the evolutionary argument is compelling for very quick decisions and we often cannot say how we arrived at a particular judgement other than say it was a 'gut feeling'. The popular classification of cognitive biases as errors brought about by heuristics - the unconscious decision-making processes in the brain - is more a matter of academic than practical interest. The important point is that better decisions can be made if we are more aware of the circumstances in which cognitive biases can occur and devise ways of countering this unhelpful behaviour. Both of these factors, bias detection and mitigation, pose serious challenges to the research community, as apparent from the limited progress so far on both accounts. However, the DECISIVe workshops have stimulated research into dealing with cognitive biases in visualization, and I hope that readers of this book will find help and inspiration in its chapters.

References

1. Ali N, Peebles D (2013) The effect of Gestalt laws of perceptual organization on the comprehension of three-variable bar and line graphs. Hum Factors 55(1):183–203
2. Argenta C, Hale CR (2015) Analyzing variation of adaptive game-based training with event sequence alignment and clustering. In: Proceedings of the third annual conference on advances in cognitive systems poster collection, p 26
3. Arnott D (1998) A taxonomy of decision biases. Monash University, School of Information Management and Systems, Caulfield
4. Baron J (2008) Thinking and deciding, 4th ed
5. BBC (2014) Horizon: how we really make decisions. http://www.imdb.com/title/tt3577924/
6. Brydges NM, Hall L (2017) A shortened protocol for assessing cognitive bias in rats. J Neurosci Methods 286:1–5
7. Bush RM (2017) Serious play: an introduction to the sirius research program. SAGE Publications, Sage, CA: Los Angeles, CA
8. Business-Insider (2013) 57 cognitive biases that screw up how we think. http://www.businessinsider.com/cognitive-biases-2013-8
9. Carter CR, Kaufmann L, Michel A (2007) Behavioral supply management: a taxonomy of judgment and decision-making biases. Int J Phys Distrib Logistics Manage 37(8):631–669
10. Cho I, Wesslen R, Karduni A, Santhanam S, Shaikh S, Dou W (2017) The anchoring effect in decision-making with visual analytics. In: Visual analytics science and technology (VAST)
11. Cooper N, Da Silva A, Powell S (2016) Teaching clinical reasoning. ABC of clinical reasoning. Wiley Blackwell, Chichester, pp 44–50
12. Correll M, Gleicher M (2014) Error bars considered harmful: exploring alternate encodings for mean and error. IEEE Trans Visual Comput Graphics 20(12):2142–2151
13. Croskerry P (2016) Our better angels and black boxes. BMJ Publishing Group Ltd and the British Association for Accident & Emergency Medicine
14. Croskerry P (2017) Cognitive and affective biases, and logical failures. Diagnosis: interpreting the shadows
15. Croskerry P, Singhal G, Mamede S (2013) Cognitive debiasing 1: origins of bias and theory of debiasing. BMJ Qual Saf 2012
16. Daron JD, Lorenz S, Wolski P, Blamey RC, Jack C (2015) Interpreting climate data visualisations to inform adaptation decisions. Clim Risk Manage 10:17–26
17. Dimara E, Bezerianos A, Dragicevic P (2017) The attraction effect in information visualization. IEEE Trans Visual Comput Graphics 23(1):471–480

18. Dunbar NE, Miller CH, Adame BJ, Elizondo J, Wilson SN, Schartel SG, Lane B, Kauffman AA, Straub S, Burgon K, et al (2013) Mitigation of cognitive bias through the use of a serious game. In: Proceedings of the games learning society annual conference
19. Edwards W, Lindman H, Savage LJ (1963) Bayesian statistical inference for psychological research. Psychol Rev 70(3):193
20. Ellis G (ed) (2014) DECISIVe 2014 : 1st workshop on dealing with cognitive biases in visualisations. IEEE VIS 2014, Paris, France. http://goo.gl/522HKh
21. Ellis G, Dix A (2015) Decision making under uncertainty in visualisation? In: IEEE VIS2015. http://nbn-resolving.de/urn:nbn:de:bsz:352-0-305305
22. Evans JSB (2008) Dual-processing accounts of reasoning, judgment, and social cognition. Annu Rev Psychol 59:255–278
23. Fendley ME (2009) Human cognitive biases and heuristics in image analysis. PhD thesis, Wright State University
24. Fiedler K, von Sydow M (2015) Heuristics and biases: beyond Tversky and Kahnemans (1974) judgment under uncertainty. In: Cognitive psychology: Revisiting the classical studies, pp 146–161
25. George JF, Duffy K, Ahuja M (2000) Countering the anchoring and adjustment bias with decision support systems. Decis Support Syst 29(2):195–206
26. Gigerenzer G (1996) On narrow norms and vague heuristics: A reply to Kahneman and Tversky
27. Gigerenzer G, Gaissmaier W (2011) Heuristic decision making. Annu Rev Psychol 62:451–482
28. Gigerenzer G, Todd PM, ABC Research Group et al (1999) Simple heuristics that make us smart. Oxford University Press, Oxford
29. Gilovich T, Griffin D (2002) Introduction-heuristics and biases: then and now. Heuristics and biases: the psychology of intuitive judgment pp 1–18
30. Graber ML, Kissam S, Payne VL, Meyer AN, Sorensen A, Lenfestey N, Tant E, Henriksen K, LaBresh K, Singh H (2012) Cognitive interventions to reduce diagnostic error: a narrative review. BMJ Qual Saf
31. Green TM, Ribarsky W, Fisher B (2008) Visual analytics for complex concepts using a human cognition model. In: IEEE symposium on visual analytics science and technology, VAST'08, 2008. IEEE, New York, pp 91–98
32. Haselton MG, Bryant GA, Wilke A, Frederick DA, Galperin A, Frankenhuis WE, Moore T (2009) Adaptive rationality: an evolutionary perspective on cognitive bias. Soc Cogn 27(5):733–763
33. Hernandez I, Preston JL (2013) Disfluency disrupts the confirmation bias. J Exp Soc Psychol 49(1):178–182
34. Heuer RJ (1999) Psychology of intelligence analysis. United States Govt Printing Office.
35. Heuer RJ, Pherson RH (2010) Structured analytic techniques for intelligence analysis. Cq Press, Washington, D.C
36. Hilbert M (2012) Toward a synthesis of cognitive biases: how noisy information processing can bias human decision making. Psychol Bull 138(2):211
37. Hogarth R (1987) Judgment and choice: the psychology of decision. Wiley, Chichester
38. IARPA (2013) Sirius program. https://www.iarpa.gov/index.php/research-programs/sirius
39. Kahneman D (2011) Thinking, fast and slow. Macmillan, New York
40. Kahneman D, Frederick S (2002) Representativeness revisited: attribute substitution in intuitive judgment. Heuristics Biases Psychol Intuitive Judgment 49:81
41. Kahneman D, Tversky A (1996) On the reality of cognitive illusions. American Psychological Association
42. Keren G, Teigen KH (2004) Yet another look at the heuristics and biases approach. Blackwell handbook of judgment and decision making pp 89–109
43. Khan A, Breslav S, Glueck M, Hornbæk K (2015) Benefits of visualization in the mammography problem. Int J Hum-Comput Stud 83:94–113
44. Kretz DR (2015) Strategies to reduce cognitive bias in intelligence analysis: can mild interventions improve analytic judgment? The University of Texas at Dallas

45. Kretz DR, Granderson CW (2013) An interdisciplinary approach to studying and improving terrorism analysis. In: 2013 IEEE international conference on intelligence and security informatics (ISI). IEEE, New York, pp 157–159
46. Kretz DR, Simpson B, Graham CJ (2012) A game-based experimental protocol for identifying and overcoming judgment biases in forensic decision analysis. In: 2012 IEEE conference on technologies for homeland Security (HST). IEEE, New York, pp 439–444
47. Manoogian J, Benson B (2017) Cognitive bias codex. https://betterhumans.coach.me/cognitive-bias-cheat-sheet-55a472476b18
48. Meehl PE (1954) Clinical versus statistical prediction: a theoretical analysis and a review of the evidence
49. Micallef L, Dragicevic P, Fekete JD (2012) Assessing the effect of visualizations on Bayesian reasoning through crowdsourcing. IEEE Trans Visual Comput Graphics 18(12):2536–2545
50. Miller S, Kirlik A, Kosorukoff A, Tsai J (2008) Supporting joint human-computer judgment under uncertainty. In: Proceedings of the human factors and ergonomics society annual meeting, vol 52. Sage, Los Angeles, pp 408–412
51. Norman G (2014) The bias in researching cognitive bias. Adv Health Sci Educ 19(3):291–295
52. Nussbaumer A, Verbert K, Hillemann EC, Bedek MA, Albert D (2016) A framework for cognitive bias detection and feedback in a visual analytics environment. In: 2016 European intelligence and security informatics conference (EISIC). IEEE, New York, pp 148–151
53. Peebles D (2008) The effect of emergent features on judgments of quantity in configural and separable displays. J Exp Psychol: Appl 14(2):85
54. Peebles D, Cheng PCH (2003) Modeling the effect of task and graphical representation on response latency in a graph reading task. Hum Factors 45(1):28–46
55. Pinker S (1990) A theory of graph comprehension. Artificial intelligence and the future of testing pp 73–126
56. Pronin E, Lin DY, Ross L (2002) The bias blind spot: perceptions of bias in self versus others. Pers Soc Psychol Bull 28(3):369–381
57. RECOBIA (2012) European Union RECOBIA project. http://www.recobia.eu
58. Remus WE, Kottemann JE (1986) Toward intelligent decision support systems: an artificially intelligent statistician. MIS Q pp 403–418
59. Roelofs S, Boleij H, Nordquist RE, van der Staay FJ (2016) Making decisions under ambiguity: judgment bias tasks for assessing emotional state in animals. Frontiers Behav Neurosci 10:119
60. Rudolph S, Savikhin A, Ebert DS (2009) Finvis: applied visual analytics for personal financial planning. In: IEEE symposium on visual analytics science and technology, 2009. VAST 2009. IEEE, New York, pp 195–202
61. Sacha D, Senaratne H, Kwon BC, Ellis G, Keim DA (2016) The role of uncertainty, awareness, and trust in visual analytics. IEEE Trans Visual Comput Graphics 22(1):240–249
62. Schlüns H, Welling H, Federici JR, Lewejohann L (2017) The glass is not yet half empty: agitation but not Varroa treatment causes cognitive bias in honey bees. Anim Cogn 20(2):233–241
63. Simon HA (1957) Models of man; social and rational. Wiley, New York
64. Soon CS, Brass M, Heinze HJ, Haynes JD (2008) Unconscious determinants of free decisions in the human brain. Nat Neurosci 11(5):543
65. Stanovich KE, West RF (2000) Individual differences in reasoning: implications for the rationality debate? Behav Brain Sci 23(5):645–665
66. Tversky A, Kahneman D (1974) Judgment under uncertainty: heuristics and biases. Science 185(4157):1124–1131
67. Tversky B (1991) Distortions in memory for visual displays. Spatial instruments and spatial displays pp 61–75
68. Valdez AC, Ziefle M, Sedlmair M (2018) Priming and anchoring effects in visualization. IEEE Trans Visual Comput Graphics 24(1):584–594
69. Verbeek E, Ferguson D, Lee C (2014) Are hungry sheep more pessimistic? the effects of food restriction on cognitive bias and the involvement of ghrelin in its regulation. Physiol Behav 123:67–75

70. Virine L, Trumper M (2007) Project decisions: the art and science. Berrett-Koehler Publishers, Oakland
71. Wall E, Blaha LM, Franklin L, Endert A (2017) Warning, bias may occur: a proposed approach to detecting cognitive bias in interactive visual analytics. In: IEEE conference on visual analytics science and technology (VAST)
72. Wichman A, Keeling LJ, Forkman B (2012) Cognitive bias and anticipatory behaviour of laying hens housed in basic and enriched pens. Appl Anim Behav Sci 140(1):62–69
73. Xiong C, van Weelden L, Franconeri S (2017) The curse of knowledge in visual data communication. In: Talk given at the information visualization research satellite event at vision sciences society annual meeting, St. Pete Beach, FL
74. Zacks J, Tversky B (1999) Bars and lines: a study of graphic communication. Memory Cogn 27(6):1073–1079
75. Zhang J, Norman DA (1994) Representations in distributed cognitive tasks. Cogn Sci 18(1):87–122
76. Zhang Y, Bellamy RK, Kellogg WA (2015) Designing information for remediating cognitive biases in decision-making. In: Proceedings of the 33rd annual ACM conference on human factors in computing systems. ACM, New York, pp 2211–2220
77. Ziemkiewicz C, Kosara R (2010) Implied dynamics in information visualization. In: Proceedings of the international conference on advanced visual interfaces. ACM, New York, pp 215–222
78. Zuk T, Carpendale S (2007) Visualization of uncertainty and reasoning. In: International symposium on smart graphics. Springer, Berlin, pp 164–177

Part I
Bias Definitions, Perspectives and Modeling

Chapter 2
Studying Biases in Visualization Research: Framework and Methods

André Calero Valdez, Martina Ziefle and Michael Sedlmair

2.1 Introduction

"Look, the earth is flat. I can see it with my own eyes." At sea-level, the curvature of the earth is too small to be perceivable to the human eye. The illusion of a flat earth is no hallucination. It is a limitation of the perceptual system. Yet, the realization that our planet is (relatively) spherical dates back to the early Greek philosophers around 600 BC. And the realization did not occur due to paying more attention to the visual impression, it came due to considering mathematical observations about the rotation of the night-sky and bodies of water, through science.

The scientific method was devised to investigate natural phenomena that are hidden from human sight, either because they were too small, too large, too fast, too slow or too rare for human perception. The human body and thus its perceptual system was crafted by evolution to enable survival of a primate in the savanna. Capabilities like objective measurement or accurate judgment of the external world are neither necessary nor helpful for survival. Being able to make decisions quickly with limited information and limited resources could make the difference between death by saber-tooth tiger or last-minute escape. Therefore, the human mind is equipped with heuristics that help decision-making with the aim of survival.

Today's world is drastically different! Yet, human perceptual and decision-making processes remain largely unchanged. People nowadays have to deal with different types of information, different amounts of information, and make much more delicate

A. Calero Valdez (✉) · M. Ziefle
Human-Computer Interaction Center, RWTH Aachen University,
Campus Boulevard 57, 52074 Aachen, Germany
e-mail: calero-valdez@comm.rwth-aachen.de

M. Ziefle
e-mail: ziefle@comm.rwth-aachen.de

M. Sedlmair
Jacobs University Bremen, Campus Ring 1, 28759 Bremen, Germany
e-mail: m.sedlmair@jacobs-university.de

© Springer Nature Switzerland AG 2018
G. Ellis (ed.), *Cognitive Biases in Visualizations*,
https://doi.org/10.1007/978-3-319-95831-6_2

13

decisions. Decisions, such as quickly detecting a critical pattern in an x-ray image, can make the difference between life and death. Decisions, such as stock-investments derived from numbers displayed on a computer screen, can influence the global economy. To gain trust in such decisions, visual inspection and communication of the underlying data and models is often an essential component. Ideally, the data and models are mapped to visual presentations that are easy to process and easy to convert into mental representations. The visual presentations should be as accurate as possible. Or as Edward Tufte put it: Keep the lie factor between 0.95 and 1.05 [34]. So in theory, accurately presenting information with respect to the visual system should yield accurate decisions.

Still, our saber-tooth-fearing minds interfere. Not only is the visual system imperfect but our cognitive system also has its pitfalls. Even when a system provides information perfectly honest, human biases might distort our view of the information and lead to imperfect or outright bad decisions. For example, a business person might invest further into a project that had already cost more than expected, as the relative prospective investment to finalize the project appears smaller than the retrospective cost of not completing the project. This *sunk cost fallacy* is the reason why many publicly funded projects cost more than previously anticipated. Nobody likes to tear down the already overpriced 80%-complete 100 million dollar airport. We might invest another 10 million dollars to complete it—and then another. Could an accurate visualization have helped the business person? Should it have overemphasized the additional costs?

The body of research on such biases is extremely large. Since Kahneman received a Nobel prize for their work on biases in 2002, research regarding biases has sprouted into all kinds of fields. From distortions in perception to distortions of complex social phenomena, the spectrum of biases is very wide. A systematic (reduced) overview of the most prominent biases can be seen in Fig. 2.1. In this figure, biases are classified in three levels of hierarchy. The first level separates the assumed high-level reasoning behind the existence of the biases. All of them are rooted in the limited perceptual and memory-related capabilities. There is either too much information available, too little meaning in our model, too little time to integrate the information, or too much information to memorize.

The second level of ordering describes strategies to cope with this reasoning. Each strategy leads to several different distortions or biases. For example, the *availability heuristic* (see Fig. 2.1 at I.a.1), describes the phenomenon that we assume things to be more frequent or important depending on how easy, or how available our mental recollection of them is [35]. It's much easier remembering the 911 attack on the World Trade Center, than a toddler drowning in a home swimming pool. This leads to a misconception. People overestimate the risk of becoming a victim of a terrorist attack and underestimate the risk of drowning in a swimming pool. Another example: The *Dunning Kruger effect* (see Fig. 2.1 at III.a.12) describes the phenomenon that people with little experience in a subject overestimate their knowledge in that subject, while people with lots of experience underestimate their knowledge: "The more I learn, the more I realize how much I don't know". The anti-vaxxer thinks he has understood the required field of medicine to evaluate the efficacy of vaccinations, while the

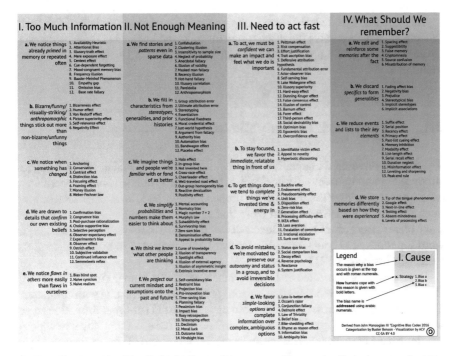

Fig. 2.1 The Cognitive Bias Codex by Buster Benson maps biases according to cause and coping strategy. This image contains hyperlinks to Wikipedia for easy look-ups. Simply click on the name of any bias to open a browser and look-up the bias.

medical researcher carefully considers different explanations and possible errors in their experimental setup. However, in such a scenario many other biases are at play.

These examples could easily benefit from visualizations depicting the real data. But, even with high-quality visualizations, biases might still persist. There is little research on biases in the field of visualization [10, 12, 36], exception for the DECISIVe workshops at VIS. Most of the aspects that have been addressed, relate to perceptual or cognitive limitations (e.g. magic number 7 ± 2) that are familiar to researchers in human-computer interaction. Other areas have been largely ignored though.

In this chapter, we draft and discuss a simple conceptual framework that can be used to guide research on biases in visualization. The framework proposes a 3-tier model of perception, action and choice, where each tier corresponds to different methods to study bias effects.

We hope that the framework will help us shed further light on the following aspects: What are interesting research questions on biases in VIS, and how can we methodologically address them? What has already happened in the cognitive sciences and what can we learn from the results and pitfalls in this large body of research?

This chapter is based on a workshop paper [6] presented the DECISIVe 2017 Workshop. It incorporates the discussions from the workshop and extends the suggestions regarding the use of certain methods for different levels of biases in our framework.

2.2 A Framework to Study Biases

The field of research on cognitive biases is large, thus organizing biases has been attempted in multiple ways. Buster Benson (see Fig. 2.1) ordered biases according to causes and strategies. Ellis and Dix [12] propose categorizing biases that occur during interpretation of visualization and those that occur later during reasoning. However some biases may occur on lower levels of perception (e.g. spider-like shapes [23] or word superiority [17]) and on higher levels of reasoning shaped by culture (e.g. the belief of a just world [21]).

Our framework is inspired by Don Norman's venerable Human Action Cycle [26]. His cycle describes seven steps that humans follow when interacting with computers. The seven steps are further classified into three stages: (1) the goal formation stage, when the user forms a goal for her/his interaction (2) the execution stage, in which a user translates the goal into actions and executes them, and (3) the evaluation stage, in which feedback from the UI is received, interpreted and compared to the user's expectations.

This model can be considered a "medium-level" model. The whole task of "perceiving" (see Fig. 2.2) is a lower-level loop on its own. On the other hand, the whole action loop in itself can be considered a sub-step in a higher-level loop model of bounded rational-choice. Naturally, these levels are not hard biological limits [1], as the cross-talk between individual steps across layers do also occur. Specifically, from a neuro-cognitive perspective, perception is less a "step-wise" open-loop procedure, but rather the convergence to a stable neural attractor state in a continuous closed-loop [1]. Perception is an active process. However, methods exist to interrupt the loop. By breaking the loop, individual steps can be studied to find step-based effects.

The idea of our framework is to provide a frame of reference when investigating a bias. In this frame of reference, different biases can occur on, or between different levels. And different methods and methodologies might be necessary to investigate biases on different levels.

Our framework differs from the categorization presented in Fig. 2.1 as it refers to different levels of cognitive processing, whereas Buster Benson's categorization follows a "cause-strategy" logic. Our framework works orthogonally to the categorization as it provides a multi-scale model of cognitive processing. While the categorization of Benson is helpful in organizing biases, it offers little insight into how to analyze, measure and counter-act a bias methodologically. Our framework aims to help in studying cognitive biases in visualization research by suggesting methods for the different levels of cognitive processing where biases may occur.

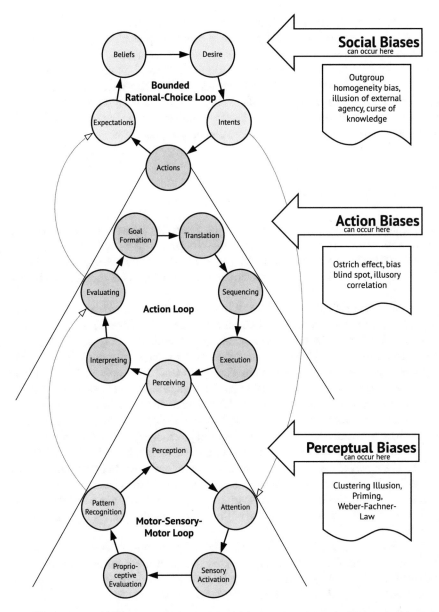

Fig. 2.2 Layered closed-loop perception, action and choice model. Since no hard boundaries exist between layers, cross-talk is part of the closed loop model (see exemplary dashed arrows)

For example, the clustering illusion is caused by a systematic tendency of the pattern recognition step in the motor-sensory-motor loop (see Fig. 2.2). This step is prone to overemphasizing possible patterns. Further, cross-talk is at play. When a person is looking for a certain pattern (i.e. bounded rational-choice step: Intent), his attention is directed towards such patterns (crosstalk). This attention pre-activates the sensory systems and in turn, leads to biased evaluation and pattern recognition. Identifying and understanding such a bias in visualization would require identifying methods to isolate the steps. Other biases can be mapped similarly.

2.2.1 Perceptual Biases

Perceptual biases refer to biases that occur on a perceptual level. In our framework, this layer is based on the motor-sensory-motor loop by Ahissar and Assa [1]. Examples of such biases are the clustering illusion, Weber-Fechner Law, or priming biases. The perception itself is biased here. One cannot "unsee" the distortion caused by the bias.

The **clustering illusion** [25, 32] is a bias that explains why people see patterns in small sets of random data. People underestimate the consequence of variance and how even little sets of random data might have clustered data. A typical example is, that if you throw a dice three times and it turns out three sixes, people will assume that the dice is unfair. And they might feel quite confident about it. However, the sequence "1–2–3" is equally probable as the sequence "6–6–6", since the throws are statistically independent. This bias is important for visualization research, as users of a visualization could over-interpret low-density scatter-plots and draw causal conclusions were non exist. Creating proper null plots, that is, visualizations showing simulated data from the null hypothesis, could be a remedy for this bias [3, 37].

The **Weber-Fechner Law** is a famous finding of early psychophysics indicating that differences between stimuli are detected on a logarithmic scale. It takes more additional millimeters of radius to discern two larger circles than two smaller circles [16]. This type of bias is probably one of the most researched biases in visualization research [15, 20].

Priming relates to findings from theories of associative memory. It refers to the idea that concepts are more quickly activated after a similar concept has been activated. The "prime" warms up the neural circuitry associated with the target, which allows faster recognition of the target. The term "so_p" is more easily completed to "soup" if you have heard terms like *butter, bread, spoon*. It's more easily recognized as "soap" when terms like *shower, water, bath* were heard before [24, 33]. This could have effects on recognizing patterns or separability in visual perception if such patterns or results have been pre-activated [7].

2.2.2 Action Biases

Action biases refer to biases made in decision-making. That is, when the perception is adequately mapped to a mental representation. Yet, the interpretation or evaluation of the percept is distorted. These biases can be reduced by training. However, even skilled people underestimate how much they are prone to biased decision-making— as stated by the Bias Blind Spot [28]. These biases occur on the second loop in our research framework—the human action loop [26].

Typical examples of action biases are the ostrich effect, illusory correlation, anchoring effects and the aforementioned availability heuristic.

The **ostrich effect** [19] describes an individual's tendency to overlook information that is psychologically uncomfortable, like the proverbial ostrich that buries his head in the sand. If you want to know why you tend to forget your full schedule, when accepting reviewer invitations: Blame the ostrich effect. It is important to study this bias in visualization research, as users might overlook information (such as a busy schedule) and make decisions not in their best interest. Visualizations aware of risks and consequences could try to compensate for such effects [11].

Illusory correlation refers to the tendency of humans to seek correlation in events that occur contingently in time [14]. Humans seek meaning in things that occur at the same time. This leads them to overestimate correlation of low-frequency events with other less familiar high-frequency events. Giving your son the name "Osama" seems inappropriate to a person inexperienced in Arabic naming frequencies, as their association with this name might be most strongly with Osama bin Laden. However, the name Timothy does not evoke such associations as it also occurs frequently in other contexts (other than the Oklahoma City Bombing by Timothy McVeigh). Illusory correlations also seem to be the reason for racial stereotyping. Such effects could be countered in a visualization by emphasizing proportions of populations. Good examples are absolute risk visualizations as Euler glyphs [5].

2.2.3 Social Biases

Social biases refer to biases that affect judgment on a social level. These effects occur on the highest level, the bounded rational-choice loop, because of cumulative effects on lower levels or because of imperfect memory. Social biases occur because of systematic biases during socialization (e.g. limited linguistic capacity implies limited cognitive capacity [29]). Famous biases in this category are the curse of knowledge, the outgroup homogeneity bias, or the illusion of external agency. Social biases should be affected by culture, while action biases should not.

The **outgroup homogeneity bias** refers to the phenomenon that people tend to see people outside their own peer group to be more homogeneous than their own in-group [27]. This is on the one hand caused by the availability heuristic—I have more memories of individual differences among my friends than among others. It is, on the other hand, also caused by imperfect memory and stereotypical memories. That's why one might believe that foreigners are all "terrorists and free-loaders" and might not be able to perceive the diversity of motivation in foreigners. It might be interesting to investigate, for example, whether labeled data in a scatterplot visualization leads to improved separability if one of the labels refers to a typical out-group and another to an in-group of the user.

The **curse of knowledge** refers to the phenomenon that once a person has acquired knowledge they may no longer be able to take the perspective of someone not having that knowledge [4]. This is why teaching or writing is hard. You, always understand what you intended to write, but everyone else might have a harder time grasping your ideas. This is also relevant for visualization research. When designing a visualization iteratively, it merges the collective knowledge of end users and developer [31]. In the end, both believe the visualization is perfectly intuitive. They might, however, overlook features that are based on their extensive knowledge from the development phase. New users might have a harder time understanding what your intricate visualization design might mean.

The **illusion of external agency** [13] refers to the illusion that the quality of an experience, that is explained to have been optimized for the recipient is rated as better than an experience without such explanation. The external agent's reality, however random it might actually be, causes a differently constructed internal reality. This is important in visualization as something that might be mistaken for a recommendation, e.g. the first item on a list, is perceived to be a better solution than any other, even if no such recommendation was ever planned. Visualizations should be careful in depicting information first if there is no intention behind this choice.

2.3 Methodological Considerations When Studying Biases

It is important to understand how biases affect judgments, specifically as visualization usually aims at providing objective information. However, studying such biases is not as easy as it might seem. Some biases might counteract each other, and experiments have to be meticulously planned to isolate the desired effect from other effects.

By identifying on which level of cognitive processing the bias occurs, it becomes simpler to pick a method to identify and measure the strength of the bias in a given scenario.

2.3.1 Perceptual Biases

Perceptual biases can be measured quite effectively using methods from psychophysics such as *staircase procedures* [8]. These procedures are designed to measure detection thresholds or just-noticeable differences between stimuli, by adaptively approaching indiscernible small differences [2].

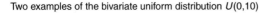

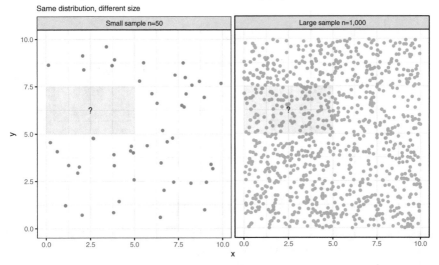

Fig. 2.3 Clustering illusion in two scatterplots. Is there a pattern in the region with the question mark?

If we take the *clustering illusion* (see Sect. 2.2.1) as an example, the bias reflects the amount of non-existing patterns detected by a user. If we assume, that such patterns are more readily detected (despite their absence) in smaller sample sizes, we contrast two scatterplots side-by-side (as seen in Fig. 2.3) where either plot does or does not contain patterns. We now ask participants, which plot shows patterns and which does not. By adjusting the sample sizes on both plots, and randomizing which of the two actually contains a pattern, we can determine the effect of sample size on detecting patterns. However, since participants could also guess correctly, we need to determine the threshold of detection using, for example, the aforementioned staircase procedure [22]. For this purpose we decrease the sample size from a starting value, by n samples (e.g. 1000 in a first step, $n = 50,950$ in a second step) until the users start detecting patterns. We then start increasing the sample size by n until the users stop detecting the patterns. At this point, we start decreasing the sample sizes again (and repeat the whole process). The mean of the inflection points determines the sample size where the bias starts working. By presenting more than two plots, we can increase certainty by reducing the probability of guessing.

If we want to investigate the effect of priming in visualizations other methods can be used. One approach aims at cutting the closed loop in the motor-sensory-motor loop [1]. This can be achieved by *subliminal activation* of primes (<100 ms) and backward masking (showing another stimulus), before the priming stimulus even reaches higher levels of cognitive processing.

A suitable example for subliminal activation to detect the effect of priming could be constructed as follows. Assuming that the previous exposure to a scatterplot with

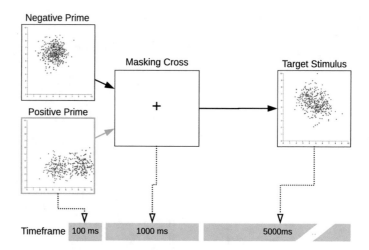

Fig. 2.4 Experimental procedure for subliminal activation. By masking the subliminal prime (either negative of positive) with a non-subliminal mask, higher levels of cognition are prevented from influencing the decision on the target stimulus

two classes, primes the separability of classes in a second visualization [7]. The higher the separability in the first plot, the easier it is to detect the separability in the second. To prevent the separability of the first plot causes an increase on the second plot by higher levels of cognition, one must prevent the first plot from reaching such levels (see Fig. 2.4). By merely exposing the first plot for less than 100 ms and immediately showing a masking stimulus for a longer period (e.g. a masking cross), the first plot is not evaluated on a higher cognitive level. Only the pre-activation of lower-level neurons helps with increasing separability in the second plot.

2.3.2 Action Biases

Methods to measure *action biases* are already far more diverse and tailored to the bias. For example, studies measuring anchoring effects explicitly try to minimize the effect, by instructing participants to disregard the anchor. The anchoring effect refers to the bias that any stimulus presented before an estimation task serves as an anchor for this estimation. For example: If we tell you the number 14 and then ask you how many species of penguins exist, your reply is going to be closer to 14 than if we had told you the number 412. In this case, your reply would be closer to 412, even if we told you this number should have no influence on the next question and instructed you to ignore it. But, how do you map such a procedure to visualization research?

An anchoring effect can indirectly be caused by letting participants derive or readnumeric outcomes from a serious of unrelated tasks. The numeric outcome

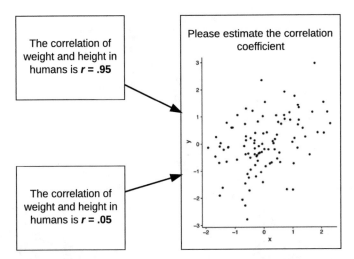

Fig. 2.5 Participants first see either the upper text or the lower text, then the target is presented. The results from the left column (either A $r = 0.95$ or B $r = 0.5$) should affect the estimation of the target question on the right (where $r = 0.4$ is correct)

should affect later evaluation due to the aforementioned *anchoring effect*. As an example, one could ask participants to first read a correlation coefficient, either large ($r = 0.95$) or small ($n = 0.05$), and then let participants estimate correlation coefficients from bivariate scatterplots. Depending on the numeric value in two different pre-conditions the correlation coefficient of the target stimulus should be shifted upwards for large anchors and downwards for smaller anchors (see Fig. 2.5). In this case, the estimated and reported correlation coefficient would be larger (e.g. $r = 0.5$ instead of the actual $r = 0.4$), if the larger correlation were shown to the participants.

2.3.3 Social Biases

If you address *social biases*, the methodology is even more dependent on the individual bias. If a bias is based on other biases, one must make an effort to estimate the biases' individual contributions to the overall effect and reduce additional systematic measurement errors. For instance, if you wanted to measure the *outgroup homogeneity bias* in a visualization, one could imagine visualizing very similar or even the same data but present it in different contexts. Similar data (i.e. same statistical properties) could be used to depict data of the participant's ingroup and in another case of the participant's outgroup. Then the participant is asked to rate the similarity of samples from the data (see Fig. 2.6). The challenge is to ensure that no lower-level biases cause a measurable bias (e.g. anchoring, priming, and pattern illusion). To

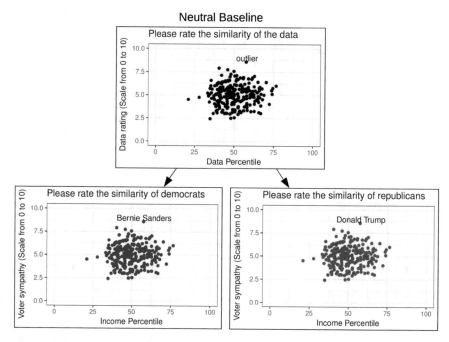

Fig. 2.6 Depending on the political orientation of the participant, different ratings of similarity are expected for the lower two visualizations, although the distributions are exactly the same. The colors are shown for comprehensibility. In a real experiment, no difference in color should be used

reduce such effects, one could run the exact same study in an abstract fashion, that is, without the labels and tasks giving away the context for the respective groups.

As an example, one could imagine comparing politicians with respect to their similarity in a given party. Voters are—if the bias is present—more likely to rate the similarity of party members, if they would traditionally vote for the opposing party (see Fig. 2.6).

2.3.4 Application of the Framework to Derive a Model

In order to utilize the framework when conceptualizing new research for a given bias, one should identify where in the framework the bias is "active". Where does the "irrational behavior" occur. Then one must aim at finding ways to minimize error from other processes in the framework.

If a bias works on the perceptual level, one must figure out a way to reduce the influence of the bounded rational-choice loop and the action loop. Mostly this is achieved, by applying methods from psychophysics research.

On the other hand, if a social bias is at work one must make sure that no lower levels of perception influence decision-making. For this reason, the data in our example task were exactly the same preventing slight differences in the pattern recognition step at a lower level. However, the choice of color could also affect the judgment.

2.3.5 Threats to Validity

In an open letter from Kahneman [18] published in *Nature*, the Nobel laureate asks researchers in the field of social priming to be cautious to publish results quickly without extensive consideration and replication. Inexperienced researchers might overlook systematic errors in experimental setups that cause distorted data indicating bias effects where none are present. The replication crisis [30] has shown that many social-psychological experiments were not reproducible. Therefore, measures to enhance reproducibility must be undertaken. It is crucial to identify methods, their benefits and pitfalls, to understand how reliable findings actually are.

To ensure that biases are measured to the highest of standards, the VIS community should also follow guidelines as presented in the open letter by Kahneman [18]. However, it makes sense to start with small setups and first gather hypotheses. The guidelines should increase reproducibility and encompass rules such as reporting confidence intervals for long-term meta-analytical research [9], open-data, preregistration of trials and publishing of negative findings. In summary, he suggests:

- Effects should be reported using *confidence intervals* to enable long-term meta-analytical cumulative research [9].
- Studies should provide all data as *open data* to allow other researchers to verify findings or even look for other explanations. If possible, release all analysis code.
- If possible, trials should be *pre-registered*, e.g. in the Open-Science-Framework.
- Sample sizes should be large enough, to ensure sufficient *statistical power* to match the expected effect size.
- Use technology to ensure all data is recorded.
- Publish *negative findings*.
- Several groups should try to validate the results of other groups. Kahneman proposed *daisy chaining*, where the results of each lab are replicated by the following lab.
- Replication should be conducted on the five most robust effects, if five groups participate in the daisy chain.
- Have guest researchers from within the daisy chain to ensure confident replication.
- Replication studies should have *larger samples* than original studies.

2.4 Conclusion

We presented a lightweight framework that helps to classify bias research in visualization. Our framework provides a frame of reference for selecting research methods when trying to identify a bias in visualization research. We believe that visual biases are a fascinating area with ample opportunities for future work. Focusing on perceptual and action biases first seems a viable road to start this process, specifically as higher-level biases are highly vulnerable to methodological flaws, apparent in the many discussions about Kahneman's famous work on "thinking fast and slow" [18]. However, carefully studying low-level perceptual and action biases will make up for a good underpinning, not only for eventually a better understanding high-level phenomena, but also as a way to better understand decision-making with visualization in general. Good practice, such as reproducibility through publishing all data, codes and experimental setup, using confidence intervals to allow for meta-analysis and reporting negative findings, will be essential in this process.

As soon as such effects are better understood in visualization, we also can start to counteract them. However, this will raise important philosophical questions: How far is it valid to correct for these biases? Challenging current views [34], should a visualization "lie" to counteract biases and improve decision-making? While for perceptual biases the answer might be quite clear, what about higher-level biases? Should the visualization decide what is in the best interest of the user? For example, may a visualization override the user's preference to not know unpleasant information and counteract the ostrich effect? A good amount of research will be needed to answer these questions and to integrate the existence of biases into visualization research.

Acknowledgements The authors wish to thank the reviewers. This work was partly funded by the German Research Council DFG excellence cluster "Integrative Production Technology in High Wage Countries", and the FFG project 845898 (VALID).

References

1. Ahissar E, Assa E (2016) Perception as a closed-loop convergence process. Elife 5(e12):830
2. Baird JC, Noma EJ (1978) Fundamentals of scaling and psychophysics. Wiley, New York
3. Beecham R, Dykes J, Meulemans W, Slingsby A, Turkay C, Wood J (2017) Map lineups: effects of spatial structure on graphical inference. IEEE Trans Visual Comput Graphics 23(1):391–400
4. Birch SA, Bloom P (2007) The curse of knowledge in reasoning about false beliefs. Psychol Sci 18(5):382–386
5. Brath R (2012) Multi-attribute glyphs on Venn and Euler diagrams to represent data and aid visual decoding. In: 3rd international workshop on euler diagrams, p 122
6. Calero Valdez A, Ziefle M, Sedlmair M (2017) A framework for studying biases in visualization research. In: Proceedings of the 2nd DECISIVe workshop 2017 held at IEEE VIS
7. Calero Valdez A, Ziefle M, Sedlmair M (2018) Priming and anchoring effects in visualization. IEEE Trans Visual Comput Graphics 24(1):584–594
8. Cornsweet TN (1962) The staircase-method in psychophysics. Am J Psychol 75(3):485–491
9. Cumming G (2012) Understanding the new statistics: effect sizes, confidence intervals, and meta-analysis. Routledge, London

10. Dimara E, Dragicevic P, Bezerianos A (2016) Accounting for availability biases in information visualization. arXiv preprint arXiv:161002857
11. Dragicevic P, Jansen Y (2014) Visualization-mediated alleviation of the planning fallacy. In: IEEE VIS 2014
12. Ellis G, Dix A (2015) Decision making under uncertainty in visualisation? In: IEEE VIS workshop on visualization for decision making under uncertainty (VDMU)
13. Gilbert DT, Brown RP, Pinel EC, Wilson TD (2000) The illusion of external agency. J Pers Soc Psychol 79(5):690
14. Hamilton DL, Gifford RK (1976) Illusory correlation in interpersonal perception: a cognitive basis of stereotypic judgments. J Exp Soc Psychol 12(4):392–407
15. Harrison L, Yang F, Franconeri S, Chang R (2014) Ranking visualizations of correlation using Weber's law. In: Proceedings of the ieee information visualization symposium (InfoVis), vol 20(12), pp 1943–1952
16. Hecht S (1924) The visual discrimination of intensity and the Weber-Fechner law. J Gen Physiol 7(2):235–267
17. Johnston JC, McClelland JL (1973) Visual factors in word perception. Attention Percept Psychophys 14(2):365–370
18. Kahneman D (2012) A proposal to deal with questions about priming effects. Nature 490
19. Karlsson N, Loewenstein G, Seppi D (2009) The ostrich effect: selective attention to information. J Risk Uncertainty 38(2):95–115
20. Kay M, Heer J (2016) Beyond Weber's law: a second look at ranking visualizations of correlation. IEEE Trans Visual Comput Graphics 22(1):469–478
21. Lerner MJ (1980) The belief in a just world. In: The Belief in a just World. Springer, Berlin, pp 9–30
22. Levitt H (1971) Transformed up-down methods in psychoacoustics. J Acoust Soc Am 49(2B):467–477
23. LoBue V (2010) And along came a spider: an attentional bias for the detection of spiders in young children and adults. J Exp Child Psychol 107(1):59–66
24. Meyer DE, Schvaneveldt RW (1971) Facilitation in recognizing pairs of words: evidence of a dependence between retrieval operations. J Exp Psychol 90(2):227
25. Morgan M, Hole GJ, Glennerster A (1990) Biases and sensitivities in geometrical illusions. Vision Res 30(11):1793–1810
26. Norman D (1988) The design of everyday things. Doubled Currency
27. Park B, Rothbart M (1992) Perception of out-group homogeneity and levels of social categorization: memory for the subordinate attributes of in-group and out-group members. J Pers Soc Psychol 42(6):1051
28. Pronin E, Lin DY, Ross L (2002) The bias blind spot: perceptions of bias in self versus others. Pers Soc Psychol Bull 28(3):369–381
29. Rice ML, Hadley PA, Alexander AL (1993) Social biases toward children with speech and language impairments: a correlative causal model of language limitations. Appl Psycholinguistics 14(4):445–471
30. Schooler JW (2014) Metascience could rescue the 'replication crisis'. Nature 515(7525):9
31. Sedlmair M, Meyer M, Munzner T (2012) Design study methodology: reflections from the trenches and the stacks. IEEE Trans Visual Comput Graphics 18(12):2431–2440
32. Seizova-Cajic T, Gillam B (2006) Biases in judgments of separation and orientation of elements belonging to different clusters. Vision Res 46(16):2525–2534
33. Strack F, Mussweiler T (1997) Explaining the enigmatic anchoring effect: mechanisms of selective accessibility. J Pers Soc Psychol 73(3):437
34. Tufte E, Graves-Morris P (2014) The visual display of quantitative information (original publish 1983)
35. Tversky A, Kahneman D (1973) Availability: a heuristic for judging frequency and probability. Cogn Psychol 5(2):207–232
36. Verbeiren T, Sakai R, Aerts J (2014) A pragmatic approach to biases in visual data analysis. In: IEEE VIS 2014
37. Wickham H, Cook D, Hofmann H, Buja A (2010) Graphical inference for infovis. IEEE Trans Visual Comput Graphics 16(6):973–979

Chapter 3
Four Perspectives on Human Bias in Visual Analytics

Emily Wall, Leslie M. Blaha, Celeste Lyn Paul, Kristin Cook and Alex Endert

3.1 Introduction

Visual analytic applications foster exploratory data analysis by combining computational techniques with interactive visualizations. A critical aspect of visual analytics is understanding how to incorporate user feedback. Such human-in-the-loop approaches to analysis allow people to leverage their domain expertise and reasoning abilities to make sense of data and gain insight. Particularly relevant to *mixed-initiative systems* [22], the principles that frame our understanding of these systems include a balance of responsibility between systems and people (i.e. an understanding of who does which specific tasks). When successful, machines and people work together to engage in a dialog about the data.

In visual analytics, we observe a trend in how user interaction is incorporated. Firstly, systems can take direct input from users to change views, direct analytic models and perform other analytic tasks. Secondly, we observe a rise in systems that learn from people's interactions and behaviors, build user models, and adapt the system based on the system's interpretation of the user's interests, analytic process,

E. Wall (✉) · A. Endert
Georgia Institute of Technology, Atlanta, GA, USA
e-mail: emilywall@gatech.edu

A. Endert
e-mail: endert@gatech.edu

L. M. Blaha · K. Cook
Pacific Northwest National Laboratory, Richland, WA, USA
e-mail: leslie.blaha@pnnl.gov

K. Cook
e-mail: kris.cook@pnnl.gov

C. L. Paul
U.S. Department of Defense, Washington, D.C., USA
e-mail: clpaul@tycho.ncsc.mil

© Springer Nature Switzerland AG 2018
G. Ellis (ed.), *Cognitive Biases in Visualizations*,
https://doi.org/10.1007/978-3-319-95831-6_3

etc. Generally, both approaches to incorporating user interaction result in people guiding the analytic process by adapting the way data is computed, visualized and otherwise transformed.

However, what if this human guidance is faulty? Whilst people have immense sensemaking and reasoning abilities as well as valuable domain expertise, we also know that people are susceptible to innate biases. In current system designs and implementations, these biases can be incorporated and propagated throughout the system. For example, if someone exhibits confirmation or anchoring bias while analyzing data, the analytic models and views could amplify the bias and lead to potentially biased recommendations or computational processes. How do we build mixed-initiative visual analytic systems that are aware of this challenge and ideally guard against it? Predominantly, current systems are agnostic to the quality and content of the guidance, operating in a reactive mode to human inputs and interactions. Recent work has begun to address how bias materializes in visual analytics [8, 9, 18, 51, 52, 54], which can point toward ways to make systems adaptive or even proactive about user biases.

An important first step toward understanding and leveraging bias is to review how we might define and formalize human bias in the scope of mixed-initiative visual analytics. Cognitive, behavioral, and social sciences have described many ways bias can occur in people's analytic processes [25, 40, 43], decision-making strategies [3, 10] and other behaviors. Motivated by the overloaded use of the term "bias" to describe different models and concepts, we thus present four perspectives on human bias, including (1) bias as a cognitive processing error, (2) bias as a filter for information, (3) bias as a preconception, and (4) bias as a model mechanism. These perspectives represent four commonly adopted takes on the term "bias". The four perspectives are not mutually exclusive; rather, they present different, potentially overlapping perspectives on bias relevant in the context of visual analytics.

To discuss how bias can affect visual analytics with a more concrete example, consider the following. Suppose Susan is using a visual analytic tool to explore possibilities for purchasing a new home. She uses the tool to browse photos, explore different areas of the city, and refine her understanding of what features of a home are important to her. From her exploration, she intends to view the homes in person and ultimately make a purchasing decision. Throughout the following sections, we will describe how each perspective on bias can impact Susan's analysis process and visual analytics in general. For each perspective, we provide a brief description, present an example scenario, and discuss how these perspectives inform and influence visual analytics.

3.2 Bias as a Cognitive Processing Error

Description: From heuristics and bias research, bias is an error resulting from an unconscious deviation from rational behavior. Cognition is frequently conceptualized as a dual-process [7]. The two processes are often termed "intuition" and "reason" [24], the former being responsible for making quick, automatic decisions and the latter being responsible for making deliberate, reflective decisions. It is one's quick judgments that are usually subject to errors.

Stanovich and West referred to the two cognitive processes as System 1 (intuition) and System 2 (reason) [45]. In this analogy, System 1 is largely subconscious and prone to making errors (bias), while System 2 is responsible for recognizing and correcting errors through intentional deliberation. These types of errors result from shortcuts in cognition, broadly referred to as heuristics [24]. Bias then is described as the method or mechanism by which the error occurs. However, the process of heuristic decision-making does not always lead to errors; it usually facilitates fast decision-making.

Example: From this perspective, there are dozens of types of bias. One such example is anchoring bias [50], which refers to the tendency to be heavily reliant on an initial value or anchor. It is analogous to a center of mass: people are unlikely to strongly deviate from their center. In Susan's home-buying scenario, she will likely be subject to anchoring bias during the price negotiation of her purchase. That is, the home's initial list price forms an anchor point and will thus subconsciously impact the amount she is willing to offer. Susan's offer for the home might have been very different had she made an offer given a different initial list price. She might even pay more money for the same home due to the tendency not to strongly deviate from the anchor point. Systems apprised of probable cognitive errors like anchoring bias have the potential to help users make better decisions by guarding against such errors, providing appropriate counterexamples, or by suggesting other ranges of data values that a user might consider.

Relevance to Visual Analytics: Common heuristic errors include confirmation bias [35], which describes the way people tend to accept confirmatory evidence of a pre-existing hypothesis and dismiss contrary information. Another common error is availability bias [49], where people tend to rely more heavily on information that is easily remembered (e.g. most recent). Similarly, the attraction effect [23] describes the tendency for a decision to be influenced by an inferior alternative. Collectively, these errors shape the way people search for and interpret information. Recently, Dimara et al. [9] showed that the attraction effect is present in users of information visualizations when interpreting scatterplots. Similarly, researchers have shown that priming and anchoring effects can be replicated in visualizations and visual analytics [8, 52]. Hence, bias impacts users outside of laboratory decision-making studies and can lead to incorrect decisions and inefficiencies in visualization-supported analytic processes.

3.3 Bias as a Filter for Information

Description: Bias acts as a filter through which we manage and perceive information. The challenge of information overload [32] motivates this analogy. Information overload, now commonly leveraged in consumer research to influence purchasing behavior, refers to a point beyond people's cognitive and perceptual limits where performance and decision-making suffer [31]. Under overload conditions, people selectively allocate attention and other mental resources to the tasks or information of highest priority. One's filter or bias thus determines what information is gathered and how sensory information is distinguished and interpreted [16].

The literature on goal-directed attention and resource allocation posits that all perception is guided by top-down influences, such as the allocation of endogenous attention [11, 41, 46]. Top-down perception governs which sensory information is identified in a scene based on goals. Bias does not make for a purely objective filter for information, however. Heuer refers to perception as an "active" process that "constructs" reality [20]; this is in contrast to a passive process that simply records reality. Similarly, obvious or important information is sometimes filtered out. For example, in one classic selective perception task, participants were shown video footage of people wearing either white or black shirts passing a basketball. Participants were asked to count how many times white-shirt basketball players on a team passed the ball to each other [44]. Most participants count the appropriate number of passes but about half fail to perceive a glaringly misfit player walk across the court. The misfit player is in a black outfit and is consequently treated as part of the task that is selectively ignored while attention is focused on the white-shirt players. In contrast to top-down perception, bottom-up perception refers to the way external factors influence attention [42]. When there is a loud noise or someone says your name across the room, you notice despite top-down attentional and perceptual focus. Visual attention can be similarly grabbed by flashing, movement or other visual cues in a display.

Example: In our home-buying scenario, Susan may experience information overload [32] as she explores homes on the market in a visual analytic tool. She might see hundreds of homes available in the area, each with dozens of attributes. Thus, her filter or bias will govern which information she perceives and which she dismisses. For example, she may only select to view single-family homes, removing condominiums, town homes, and apartments from the visualization. If removed, some options that may be relevant to Susan's other search criteria will not be visibly available, though still in the underlying data and system. The system may want to make some of that information known at an appropriate point in the analytic process. By leveraging knowledge about people's perceptual strengths and limitations, a mixed-initiative system could present information in ways that are easy for users to understand and at a time when mental resources are available.

Relevance to Visual Analytics: A great deal of research in perception has been leveraged by researchers in information visualization and visual analytics to present information in ways that are most perceptually accessible [13]. Pre-attentive processing theory [47], for example, describes the nature and limits of visual information processing. In creating visual representations of data, this is often used by designers as a guide to prevent overwhelming a user's perceptual limitations. Similarly, Gestalt principles [26] refer to the relationships inferred by the visual system based on proximity, groupings, symmetry, etc. between visual elements. Thus, understanding how people's filters work can inform things like which visual widgets or elements to place in close proximity to one another or which graph layout algorithm is most appropriate. Indeed, Patterson et al. [39] listed supporting attention and user mental models as some of the key visualization leverage points for design grounded in human cognition.

3.4 Bias as a Preconception

Description: Analysts approach mixed-initiative systems bringing all their experiences and internal influences that unconsciously shape their approaches to the analysis process. This, in turn, influences the ways they interact with systems. The consequence is that the user model within the system, the analytic products, and provenance may be shaped by each individual's unconscious biases. These types of bias may seem to have little to do directly with the task at hand. Yet, because they shape the person, there is a high likelihood they can influence mixed-initiative sensemaking.

Unconscious biases arise in a number of ways. They derive from a person's cultural beliefs and traditions, which include their implicit assumptions and expectations regarding stereotypes. Unconscious biases result from general self-confidence or self-esteem, as well as comfort or familiarity level with the capabilities of a machine's analytics and interface functions. Related personality traits render some people more risk seeking or risk averse, shaping how they push boundaries exploring a space of hypotheses or push the capabilities of the computational system. These characteristics are thus seen as a source of individual variability between people.

Example: Susan is avoiding listings for houses downtown in the city. Having lived in the suburbs for many years, Susan assumes that neighborhoods near downtown have higher crime rates and lower economic stability. She believes she should not make a housing investment there. The availability of recent census results and police reports within the real estate analytic tools enable Susan to explore her assumptions and refine her thinking. A mixed-initiative system may detect her avoidance of downtown properties and could prompt her to challenge her assumptions with the related data.

Relevance to Visual Analytics: Unconscious biases shape analysts' assumptions and stereotypes about analytical tools and mixed-initiative aids, and they shape assumptions and stereotypes about the data/analytical subjects (e.g. presumed reliability or trustworthiness of certain sources). Implicit attitudes shape the formulation of hypotheses and the questions about the assumptions and the consequences of those hypotheses. Klein et al. posited that the entire sensemaking process begins with a practitioner framing the problem and the selected framework, however minimal, then shapes what an analyst thinks about and what structure they think with [25]. Frames reflect a perspective an analyst takes to make sense of data or to solve a problem. As implicit attitudes shape an analyst's perspective, they shape the analyst's frames, thereby shaping the sensemaking process.

Use of the system is also influenced by the level of trust the user places in computational systems, which is shaped by the degree of machine autonomy the system has together with its transparency about its capabilities and uncertainty [27]. Some people are more pre-disposed to trust computational systems. This would manifest in differences in the degree of reliance an analyst places on the machine's results or recommendations. Generally, the strategy for addressing differences in reliance and trust is to either find a means of trust calibration or to help the user adjust expectations about machine capabilities [21]. It is possible that the preconceived biases that might play into the analytic process could influence trust and reliance on the visual analytic system. Consequently, the mixed-initiative interface should be providing cues to enable the user to calibrate her/his trust in the machine as well.

Expertise, derived from general experience as well as explicit training, further shapes the analytical process and is shaped by implicit biases. Expertise can impact expectations and perceptions of a mixed-initiative system and the interpretations of the information visualizations under consideration. Expertise in forensic analytics, for example, may make analysts more conservative in their judgments, shaped in part by their expert understanding of the consequences of their decisions. Often, expertise also provides the user with a better understanding of the limitations of the analytical tools or data collection practices, which can shape more nuanced interpretations during the analysis process.

Because they are built to record a number of different types of user behaviors throughout the analysis process, mixed-initiative systems may be particularly well-positioned to aid in the assessment of unconscious biases of analysts. We argue that it is possible for a mixed-initiative system to capture and integrate unconscious, preconception biases into analytics through the user model, the systems model of the users interest, and track those biases through user interactions and changes in the user's mental model over time.

3.5   Bias as a Model Mechanism

Description: Bias is the term often used in cognitive modeling to describe a decision boundary or a tendency toward one response option over another. Cognitive models are mathematical and computational approaches to formally describe mechanisms supporting perception, memory, decision-making, and other cognitive functions [5]. A number of these models include a mechanism explicitly called bias, or they use a combination of mechanisms to capture the ways the aforementioned types of bias manifest in measurable behaviors, like response choice and speed. Models with explicit bias mechanisms often contain a bias parameter or measure bias as a relationship between parameters. Here, we will review two major perspectives on bias as a model mechanism, one which formalizes bias within models of mental organization and another which formalizes bias in models of decision-making dynamics. Both types of behavior are necessary in visual analytics, as analysts work through their sensemaking processes of organizing information and weighing evidence against potential hypotheses and interpretations. As interactive visual analytic systems aid the externalization of analysts' mental models, model mechanisms can help us interpret how bias is reflected in the patterns and dynamics of their interactions.

One approach to modeling bias addresses the question: where do people mentally "draw the line" between one response option and another when performing an analytic task? Many models of perceptual choice or organization describe information representation with two mechanisms. One mechanism is spatial organization that groups pieces of information by similarity/proximity; alike objects are close in space or clustered together. The second mechanism is at least one boundary that divides the space into response regions; object labels or choices are made according to the response regions defined by the boundary. Examples of these models include the theory of signal detection for finding signals in noise [19, 30] or categorization models [29, 37] for clustering and labeling groups of objects. Bias in these models is described by a weighting of boundary regions; if regions are not equally weighted, the model represents bias toward certain responses. Other models might capture bias as a feature weighting, representing how much the respondent emphasized certain features over others.

Another major use of bias parameters is found in models of information processing dynamics behind the time to make a decision. These dynamic decision models characterize the choice between two options as a stochastic process whereby information about the options is incrementally sampled and accumulated, often in a random walk fashion, until some threshold is reached for one of the response options [6]. The evidence accumulation process governs a person's response speed and is influenced by the salience and complexity of the choice options. Bias in these models is captured by the relationship between the starting value of the evidence accumulators and the response thresholds. If the accumulator starts at zero, then the process is not biased; all responses are equally likely. If the bias parameter is non-zero, then the process is biased toward the response threshold closer to the bias value. This bias mechanism captures behaviors wherein some responses, correct or erroneous, are selected more frequently or more quickly than others.

Example: Homes for sale are described by a large number of attributes drawn from real estate descriptions. Susan is likely to have certain features along which she is organizing the options available on the market, such as the number of bedrooms, number of bathrooms, basement square footage and proximity to schools. This forms a four-dimensional mental representation space into which the houses can be organized. If she is weighing numbers of bedrooms and bathrooms equally, we can describe her decision bias as equidistant from the category centroids or close to zero. However, Susan has strong opinions about basement square footage and proximity to schools. Based on how she organizes houses into desirable and undesirable categories, we might use models to infer that she is biased toward liking houses that are within a 10 minute walk to schools but have small basements less than 400 square feet. A system aware of these preferences might help quickly reorganize large amounts of data into a representation consistent with the user's mental representation.

Relevance to Visual Analytics: Visual analytic systems designed to support data exploration capture an externalization of the analyst's mental organization in the form of interaction. By leveraging analytic provenance [36], researchers can better understand users' strategies [10], processes that led to insights [17], and ultimately better support the sensemaking process [55]. Different spatial layouts and data encodings (including colors, shapes, etc.) reflect mental organization patterns, including the perception of similarity between data points. Characterizing the biases in this mental organization process provides a quantifiable way to describe the information representation space and decision boundaries. For example, we can use the perceptual organization models to infer if the analyst is biased toward some data attributes or certain clusters/labels. We could use the sequential sampling model to identify biases in how analysts are weighing the relative utility or value of a piece of evidence.

From the perspective that bias is a model mechanism, we can also formally characterize bias from the other three perspectives described in Sects. 3.2, 3.3 and 3.4. Although these models are implemented in a way that is rather agnostic to errors in reasoning, the bias parameters enable inferences about how errors from decision heuristics occur. For example, anchoring bias would be captured as a bias toward one of the response thresholds close to the anchor value in a model of information accumulation or decision dynamics. Bias as a filter can be formalized as a bias node or parameter in a neural network or hierarchical model of vision [48]. This would reflect the way information might be differently sampled by an analyst based on the goal-related task they are performing. Preconception bias can be included in models as latent factors or correlates of measurable behaviors. As latent factors, biases such as gender or race stereotypes can modulate other mechanisms in the mental models, such as the organization of similar objects or response preferences [53].

3.6　Discussion

These four perspectives of bias illustrate the diversity in how people process information and form a model of the world. Each is a valid perspective that greatly shapes how bias is framed in visual analytics research. However, the multiplicity of definitions sometimes leads to challenges in sharing and collaboration due to a lack of common ground. One goal of this chapter is to present these definitions, so that we as a community have a starting point for discussing how these perspectives fit within the visual analytics research agenda. Additionally, when considering all of these perspectives, the space in which to study bias in visual analytics increases dramatically. This leads to several open challenges and opportunities for the visual analytics community.

3.6.1 Does Bias Endanger Mixed-Initiative Visual Analytics?

Visual analytic applications continue to model users and adapt interfaces, visualizations, and analytic models based on user's interactions. However, how do such systems differentiate between valuable subject matter expertise (which should be incorporated) and biased input? Without such techniques for identifying and guarding against biased input, applications run the risk of showing users biased views of their data that correspond to what they want to use, rather than truthful representations of the information.

For example, in model-steering situations, user input guides analytic models to focus on salient aspects of the domain being studied [12]. Without guarding against potentially biased user input, the system may overfit the model to the biased input. The result may be a system that shows users the views they want to see, but is essentially an "echo chamber" for their own biases.

A recent example that showcases the potential consequences of human bias in systems is the AI chatbot, Tay [1, 28]. The artificial intelligence was intended to be a friendly chatbot that appealed to young adults. The underlying model was continually trained by incoming tweets, causing Tay to tweet increasingly racist and misogynistic messages shortly after going live. While a vulnerability in Tay was exploited, the chatbot nonetheless conveys what can happen when human bias is introduced unchecked into a system. An awareness of these potential risks will help us develop better systems and ultimately foster better data-driven decisions.

One approach for making the distinction between valuable domain expertise and biased input might be to consider the consistency or inconsistency of a user's interaction sequences. More sophisticated approaches could be derived by studying the differences in interaction sequences of domain experts and novices who are biased. It may also be useful to study large groups of users, expert or novice, modeling their processes and biases, to provide additional context to the machine intelligence about ranges of typical and outlier behaviors.

3.6.2 How to Keep the Machine "Above the Bias"?

Designing mixed-initiative visual analytic systems to reduce negative effects of biased user input is an interesting and important line of research leveraging our bias classifications. As noted by Friedman et al., there are three types of bias that can influence computer systems: pre-existing, technical and emergent biases [14, 15]. Pre-existing bias arises from the attitudes or societal norms/practices that the software designers might impart into system designs. This is akin to our bias as a preconception perspective. Concerted efforts can be made to address pre-existing bias throughout the visual analytics design process, such as using the recent GenderMag method to address gender biases in interface designs [4].

Technical biases are a consequence of technical considerations, such as choice of hardware or algorithm. Computational technical biases are unique from the various definitions of human bias we summarized herein. However, because they will contribute to biases in mixed-initiative system performance, careful technical choices should be made and appropriate details should be made available to the user to facilitate informed interpretation of system behaviors.

Emergent biases arise from the use of a system, resulting from changing context or knowledge in which a system is being used. Friedman argues that these are more difficult to know in advance or even identify in practice [15]. Emergent biases are highly likely to occur in mixed-initiative systems, particularly as the interface or algorithms are shaped by any of the aforementioned biases that are influencing the user's interactions. Theoretically, the role of the machine is to be unbiased and to present a rational result based on clear rules. However, there are limitations to this approach, namely the lack of tacit knowledge and analytic context that cannot be easily modeled. This has led to the rise of user-driven machine learning that goes beyond a "supervisory" role in training [2]. Yet, as soon as the human is re-introduced into the system, the rationality of the machine is affected. How can we judge when this human-machine teaming is succeeding or failing?

We propose that mixed-initiative systems are uniquely suited to aid in the identification and mitigation of emergent biases, exactly because mixed-initiative systems reflect the user's analytic process. To do this, we must be able to correctly interpret the user's biases as they are captured by the computational system. The four perspectives we have outlined will help the bias interpretation process. Each provides a way to identify how that source of bias plays out in the analytic process. To the degree that formal models are available for each bias perspective, those can be integrated into the system for more automated interpretations.

3.6.3   Could the Mixed-Initiative System Impart Bias to the User?

Yes. A less-emphasized aspect of emergent bias is that the structure of the user interface may influence and bias the interactions of the user. Reliance on machine automation and automated decision aids can result in automation bias. This is the heuristic use of automation instead of more vigilant information seeking and decision-making [33, 34, 38]. The errors resulting from automation bias are of concern for mixed-initiative systems, wherein those errors might be integrated into the analytic results/visualizations or even the analytic processes. Of particular concern in this domain are automation commission errors. These errors are inappropriate actions resulting from over-attending to automated aids without attention to the context or other critical environmental information sources. Commission errors occur when a user accepts the recommendation of some machine analytics even when there is contrary evidence from other information sources, either internal or external to the analytics system.[1] The design of an interactive analytic interface may lend itself to overemphasizing some analytic results or mixed-initiative recommendations, such as highlighting recommendations or altering things like the size or color that might make some recommendations stand out over others. Automation bias in accepting the most strongly emphasized recommendations could lead the analyst down a biased analysis path. Does the system or the user bear the responsibility for mitigating automation bias? We argue that if mixed-initiative systems can cultivate emergent biases in both the machines and the users, then mixed-initiative systems also offer new opportunities for humans and machines to team up to mitigate negative effects of bias.

3.6.4 Is Bias Good or Bad?

The term bias tends to carry a negative connotation. It is perceived as something that we should strive to eradicate. However, bias is not always bad. Each of the four perspectives on bias differs in how it impacts cognitive and perceptual processes.

From the perspective that bias is an error, we should work to minimize it; however, it should not be confused with the heuristic decision-making processes that lead to such biases. We emphasize that heuristic decision-making is not inherently bad. It usually results in more efficient decision-making. Thus, it is imperative that in attempting to mitigate bias as an error, we do not unduly limit heuristic decision-making processes in general.

From the perspective that bias is a model mechanism, it is neither good nor bad. In this case, it is an objective characterization of the decision-making process. Whilst the decision-making process itself may be suboptimal or erroneous (as is the case of bias as an error), here bias just describes the boundary between response options.

From the perspective that bias is a filter and the perspective that bias is a preconception, it can be both beneficial and detrimental depending on circumstances. Perceptual filters prevent us from experiencing information overload, however, they can also cause us to inadvertently filter out information relevant to a given decision. Unconscious biases like innate risk-aversion tendencies can help us to make deliberate, mindful decisions, but on the other side of the spectrum can lead to impulsive high-risk decisions. Thus, because different perspectives on bias vary widely in their potential benefits or risks, it is imperative to thoughtfully define the perspective and scope considered for bias detection or mitigation efforts.

---

[1]Commission errors are contrasted with automation omission errors, which occur if the human-machine team fails to respond to system irregularities or the system fails to provide an indicator of a problematic state. In visual analytics, an omission error could occur if a system "knows" an algorithm might be mis-matched to a data type but does not alert the analyst.

3.7 Conclusion

Bias is a particularly important consideration in the design of mixed-initiative visual analytic systems, where biased human input can shape analytical models. Thus, in this chapter, we have described four perspectives on bias particularly relevant to such human-machine collaborative systems. We hope that by discussing and differentiating these perspectives on the overloaded term "bias," researchers and developers can thoughtfully define which perspective they take in their work on bias.

Acknowledgements The research described in this document was sponsored by the U.S. Department of Defense. The views and conclusions contained in this document are those of the authors and should not be interpreted as representing the official policies, either expressed or implied, of the U.S. Government.

References

1. Alaieri F, Vellino A (2016) Ethical decision making in robots: autonomy, trust and responsibility. In: Agah A, Cabibihan JJ, Howard AM, Salichs MA, He H (eds) Social robotics: 8th international conference. Springer International Publishing, Kansas City, MO, pp 159–168
2. Amershi S, Cakmak M, Knox WB, Kulesza T (2014) Power to the people: the role of humans in interactive machine learning. AI Mag 35(4):105–120
3. Brown ET, Ottley A, Zhao H, Lin Q, Souvenir R, Endert A, Chang R (2014) Finding Waldo: learning about users from their interactions. IEEE Trans Visual Comput Graphics 20(12):1663–1672
4. Burnett M, Stumpf S, Macbeth J, Makri S, Beckwith L, Kwan I, Peters A, Jernigan W (2016) GenderMag: a method for evaluating software's gender inclusiveness. Interact Comput 28(6):760–787
5. Busemeyer JR, Diederich A (2010) Cognitive modeling. Sage, Los Angeles, CA
6. Busemeyer JR, Townsend JT (1993) Decision field theory: a dynamic-cognitive approach to decision making in an uncertain environment. Psychol Rev 100(3):432–459
7. Chaiken S, Trope Y (1999) Dual-process theories in social psychology. Guilford Press, New York
8. Cho I, Wesslen R, Karduni A, Santhanam S, Shaikh S, Dou W (2017) The anchoring effect in decision-making with visual analytics. In: IEEE conference on visual analytics science and technology (VAST)
9. Dimara E, Bezerianos A, Dragicevic P (2017) The attraction effect in information visualization. IEEE Trans Visual Comput Graphics 23(1):471–480
10. Dou W, Jeong DH, Stukes F, Ribarsky W, Lipford HR, Chang R (2009) Recovering reasoning process from user interactions. IEEE Comput Graphics Appl pp 52–61. http://citeseerx.ist.psu.edu/viewdoc/download?doi=10.1.1.157.407&rep=rep1&type=pdf
11. Egeth HE, Yantis S (1997) Visual attention: control, representation, and time course. Annu Rev Psychol 48(1):269–297
12. Endert A, Ribarsky W, Turkay C, Wong B, Nabney I, Blanco ID, Rossi F (2017) The state of the art in integrating machine learning into visual analytics. In: Computer graphics forum. Wiley Online Library
13. Fekete JD, Van Wijk J, Stasko J, North C (2008) The value of information visualization. Inf Visual pp 1–18
14. Friedman B (1996) Value-sensitive design. Interactions 3(6):16–23

15. Friedman B, Nissenbaum H (1996) Bias in computer systems. ACM Trans Inf Syst (TOIS) 14(3):330–347
16. Frisby JP, Stone JV (2010) Seeing: the computational approach to biological vision. The MIT Press, Cambridge, MA
17. Gotz D, Zhou MX (2009) Characterizing users' visual analytic activity for insight provenance. Inf Visual 8(1):42–55
18. Gotz D, Sun S, Cao N (2016) Adaptive contextualization: combating bias during high-dimensional visualization and data selection. In: Proceedings of the 21st international conference on intelligent user interfaces - IUI '16 pp 85–95. http://dl.acm.org/citation.cfm?doid=2856767.2856779
19. Green DM, Birdsall TG, Tanner WP Jr (1957) Signal detection as a function of signal intensity and duration. J Acoust Soc Am 29(4):523–531
20. Heuer Jr RJ (1999) Psychology of intelligence analysis. Washington, D.C
21. Hoffman RR, Johnson M, Bradshaw JM, Underbrink A (2013) Trust in automation. IEEE Intell Syst 28(1):84–88
22. Horvitz E (1999) Principles of mixed-initiative user interfaces. In: Proceedings of the SIGCHI conference on human factors in computing systems pp 159–166
23. Huber J, Payne JW, Puto C (1982) Adding asymmetrically dominated alternatives: violations of regularity and the similarity hypothesis. J Consum Res 9(1):90–98
24. Kahneman D, Frederick S (2005) A model of heuristic judgment. The Cambridge handbook of thinking and reasoning pp 267–294
25. Klein G, Moon B, Hoffman RR (2006) Making sense of sensemaking 2: a macrocognitive model. IEEE Intell Syst 21(5):88–92
26. Koffka K (2013) Principles of gestalt psychology, vol 44. Routledge, London
27. Lee JD, See KA (2004) Trust in automation: designing for appropriate reliance. Hum Factors 46(1):50–80
28. Lee P (2016) Learning from Tay's introduction. https://blogs.microsoft.com/blog/2016/03/25/learning-tays-introduction/
29. Luce RD (1977) The choice axiom after twenty years. J Math Psychol 15(3):215–233
30. Macmillan NA, Creelman CD (2004) Detection theory: a user's guide. Psychology Press, New York
31. Malhotra NK (1982) Information load and consumer decision making. J Consum Res 8(4):419–430
32. Milord JT, Perry RP (1977) A methodological study of overloadx. J Gen Psychol 97(1):131–137
33. Mosier KL, Skitka LJ (1996) Human decision makers and automated decision aids: made for each other. In: Parasuraman R, Mouloua M (eds) Automation and human performance: theory and applications. Lawrence Erlbaum Associates, Mahwah, NJ, pp 201–220
34. Mosier KL, Skitka LJ (1999) Automation use and automation bias. In: Proceedings of the human factors and ergonomics society annual meeting, vol 43. Sage, Beverley Hills, pp 344–348
35. Nickerson RS (1998) Confirmation bias: a ubiquitous phenomenon in many guises. Rev Gen Psychol 2(2):175–220
36. North C, May R, Chang R, Pike B, Endert A, Fink GA, Dou W (2011) Analytic provenance: process+interaction+insight. In: 29th annual CHI conference on human factors in computing systems, CHI 2011 pp 33–36
37. Nosofsky RM (1991) Stimulus bias, asymmetric similarity, and classification. Cogn Psychol 23(1):94–140
38. Parasuraman R, Manzey DH (2010) Complacency and bias in human use of automation: an attentional integration. Hum Factors 52:381–410
39. Patterson RE, Blaha LM, Grinstein GG, Liggett KK, Kaveney DE, Sheldon KC, Havig PR, Moore JA (2014) A human cognition framework for information visualization. Comput Graphics 42:42–58

40. Pirolli P, Card S (2005) The sensemaking process and leverage points for analyst technology as identified through cognitive task analysis. In: Proceedings of international conference on intelligence analysis 2005, pp 2–4. http://scholar.google.com/scholar?hl=en&btnG=Search&q=intitle:The+Sensemaking+Process+and+Leverage+Points+for+Analyst+Technology+as+Identified+Through+Cognitive+Task+Analysis#0
41. Posner MI (1980) Orienting of attention. Q J Exp Psychol 32(1):3–25
42. Riesenhuber M, Poggio T (1999) Hierarchical models of object recognition in cortex. Nat Neurosci 2(11):1019–1025
43. Sacha D, Stoffel A, Stoffel F, Kwon BC, Ellis G, Keim DA (2014) Knowledge generation model for visual analytics. IEEE Trans Visual Comput Graphics 20(12):1604–1613
44. Simons DJ, Chabris CF (1999) Gorillas in our midst: sustained inattentional blindness for dynamic events. Perception 28(9):1059–1074
45. Stanovich KE, West RF (2000) Advancing the rationality debate. Behav Brain Sci 23(5):701–717
46. Torralba A, Oliva A, Castelhano MS, Henderson JM (2006) Contextual guidance of eye movements and attention in real-world scenes: the role of global features in object search. Psychol Rev 113(4):766–786
47. Treisman A (1985) Preattentive processing in vision. Comput Vis Graphics Image Process 31(2):156–177
48. Tsotsos JK (2011) A computational perspective on visual attention. MIT Press, Cambridge, MA
49. Tversky A, Kahneman D (1973) Availability: a heuristic for judging frequency and probability. Cogn Psychol 5(2):207–232
50. Tversky A, Kahneman D (1974) Judgment under uncertainty: heuristics and biases. Science 185:1124–1131
51. Valdez AC, Ziefle M, Sedlmair M (2018a) A framework for studying biases in visualization research. In: Ellis G (ed) Cognitive biases in visualizations, Chap. 2. Springer, Berlin
52. Valdez AC, Ziefle M, Sedlmair M (2018b) Priming and anchoring effects in visualization. IEEE Trans Visual Comput Graphics 24(1):584–594
53. Vandekerckhove J (2014) A cognitive latent variable model for the simultaneous analysis of behavioral and personality data. J Math Psychol 60:58–71
54. Wall E, Blaha LM, Franklin L, Endert A (2017) Warning, bias may occur: a proposed approach to detecting cognitive bias in interactive visual analytics. In: IEEE conference on visual analytics science and technology (VAST)
55. Xu K, Attfield S, Jankun-Kelly T, Wheat A, Nguyen PH, Selvaraj N (2015) Analytic provenance for sensemaking: a research agenda. IEEE Comput Graphics Appl 35(3):56–64

Chapter 4
Bias by Default?

A Means for A Priori Interface Measurement

Joseph A. Cottam and Leslie M. Blaha

4.1 Introduction

Does all data in an application have an equal chance of being seen? The answer to this question is likely "no", and that is not necessarily a bad thing. We deliberately influence what is visible and what is not based on many goals or intentions with an information representation. In fact, we rely on such imbalances as part of the data exploration process to keep the information content tractable for human memory and reasoning [16]. Any time something "just pops out" or is "obvious" in a display, there is an element of bias at play. However, does the interface naturally bias in a way that supports or impedes the tasks it was designed to support? How much of that bias is inherent in the interface, and how much is the result of the ways the interface interacts with a specific dataset? How much is the result of the user crafting the interface for personal needs and interests? This chapter proposes Markov modeling as an approach to begin teasing apart the sources of bias in visual analytic systems.

Friedman and colleagues defined bias in computers systems as a slant which produces systematic and unfair discrimination against certain individuals or groups, particularly when that discrimination is paired with unfair outcomes [9, 10]. They defined three types of biases: pre-existing, technical and emergent. Although we disagree that bias only produces unfair outcomes, we find these classes useful for thinking about the bias that can estimated about the system with and without user interactions. Pre-existing bias reflects how a system embodies cultural norms, practices and attitudes that exist in the environment in which the system was developed, programmed or deployed. Pre-existing biases in visual analytics might reflect the culture of the company or research group that developed the system. They could be

J. A. Cottam (✉) · L. M. Blaha
Pacific Northwest National Laboratory, 902 Battelle Blvd, Richland, WA 99354, USA
e-mail: joseph.cottam@pnnl.gov

L. M. Blaha
e-mail: leslie.blaha@pnnl.gov

© Springer Nature Switzerland AG 2018
G. Ellis (ed.), *Cognitive Biases in Visualizations*,
https://doi.org/10.1007/978-3-319-95831-6_4

as simple as the default interface elements, like default color schemes or variable placements on axes. For example, the www.SmartMoney.com *Map of the Market* was a popular example of a treemap compactly representing stock market values [20]. The default color scheme for the map was on a red-green spectrum, with green representing positive trending stocks (gains) and red representing negative trending stocks (losses). The red-green spectrum plays off the cultural norm of green for go and red for stop, adopted from traffic signals. The treemap offered an alternative yellow-blue color map, particularly as an alternative for people with red-green color blindness; however, we have no compelling a priori cultural association for whether yellow or blue should be assigned to gains or losses. Without the pre-existing bias, we lose some intuition for interpreting the visualization.

Other system biases are technical biases. These arise from technical constraints or considerations in the design process, such as choice of hardware or peripherals, which shape the capabilities of the system. Technical bias in visual analytic systems can influence the initial layout, the available algorithms or the options for interaction techniques. Interaction options have implications for the amount of information that needs to be available on the screen. For example, hover and roll-over functions may not be enabled without a mouse or touchpad. Without a hover option, tooltips may not be possible, so information that might have been available on demand may need to be readily available in other ways or on the screen at all times. Or the burden can be placed on the user to query for the information; however, if the user is inexperienced with the system or poor at formulating queries, then some information may not be queried and so may not be seen. Another form of technical bias can be seen in the specific algorithms provided in a tool. They are often chosen based on expected performance on reference hardware for anticipated datasets. As hardware advances, previously intractable algorithms can be implemented, and as new datasets are approached with a tool, different algorithms may be preferred.

A third class of system biases are emergent biases that result from the interactions of users with the system. These are very much of interest to visual analytic systems which are meant to facilitate extensive interactions for data exploration [3]. However, we suspect that emergent biases can only be measured from user interactions with the system. This is because each user has unique biases from attitudes, experience and task goals that will shape the emergent biases [17, 18]. Whether the goal of measurement is online or post hoc bias assessment, it is hard to predict emergent biases in the absence of specific user characteristics and interaction behavior data.

Thus, the goal of this chapter is to propose a framework by which we can measure the biases of an interface from the design of the system, including choices of visualizations and interactions. This may include elements of both technical and pre-existing biases, which do not require the collection of user interaction data for assessment. Of particular interest, at present, is predicting if the system design will steer users into system states where information is systematically unavailable or hard to recover, which will bias their exploratory reasoning and inference processes. Identifying the biases a priori helps (1) identify when and which biases are important, (2) compensate for biases when they hinder task performance and (3) constructively employ biases when they help.

The chapter is structured as follows. First, we review how our concept for a priori bias compares to analytic provenance and interaction sequences. Then we overview the Gapminder tools for the Gapminder World data,[1] which we use as a running example to demonstrate measuring a priori bias. Then we introduce Markov modeling for capturing interfaces as a state space model and introduce two models for a priori system bias. Finally, we present an analysis of Gapminder visualizations. We end with discussions of other potential formalisms and next steps in using this approach to capture biases in analytic tools.

4.2 Relationship to Analytic Provenance

Modeling a priori system bias provides an important complement to analytic provenance modeling. The goal of provenance modeling is to leverage the sequence of user actions to characterize a user's analytic process [13, 21]. Xu et al. [21] argue that there are two important uses for analytic provenance: users can plan further analyses and systems can suggest related but unexamined data. If captured and interpreted automatically, rather than through intensive manual annotation, a mixed-initiative system could incorporate analytic provenance into intelligent recommendations, as illustrated by Endert et al. [6] and Cook et al. [2]. Notably, Dabek and Caban [4] use captured actions to automatically build Markov-model-like automata that form the basis of their intelligent recommender system.

Additionally, when used post hoc, provenance enables analysts to study their own and others' processes. Toward this end, there have been efforts to develop visualizations for showing analytic provenance. GraphTrail [5] uses a graph visualization approach where the states of the analytic system are nodes, and the links illustrate the analyst's transition path between the visualizations. Those links could be enriched by identifying the types of actions they represent in the analytic process, using the catalog of activity developed by Gotz and Zhou [13], for example.

From a system design perspective, analytic provenance analysis allows designers to inspect how design choices and interface elements were used throughout task completion. Our proposed Markov chain model for interface and exploration biases offers a predictive analysis for what might happen. This analysis can be conducted before the system is given to users; it can be engaged early and often in the design process. Importantly, our proposed interface and exploration bias computations are common across users, because they are about the system structure, not the specific user interactions or tasks. Thus, the emergent system biases introduced by the user interactions with the system may be teased apart from the other system biases by leveraging a combination of Markov-chain-based interface analyses and analytic provenance modeling.

Analytic provenance can then capture what a user *actually* does with a system, which can be compared to the predicted provenance from the Markov chain. We

---

[1] http://gapminder.org.

propose that modeling the system independent of user interactions is also valuable. Such modeling targets the potential biases in the system that would influence the ways a user *could* or *should* use the system. In many ways, this may be considered task-independent modeling of the potential interaction sequences. Yet, from the perspective of pre-existing bias, this process is also capturing the way system structure and readily available interactions contribute to all tasks attempted with the system. Technical or pre-existing biases may create some systems states that are not useful or would strongly sway the analytic process. While we can observe if or when analysts navigate into those states using analytic provenance, a priori modeling may help us to predict or prevent states unhelpful to the sensemaking process, or that might be compounded by user biases to create strong emergent biases.

4.3 Gapminder

Throughout the rest of this chapter, we will use the Gapminder tools as example visual analytic interfaces. Gapminder is a Swedish organization that curates data and statistics about the world, made available for research and education purposes on http://www.gapminder.org. The Gapminder World data includes variables like the population size, income per capita and life expectancy. The organization offers a set of web browser-based interactive visualizations for exploring the Gapminder World data. Figure 4.1 shows an example of the Maps visualization, which has data points plotted as color circles overlaid on the map of the world, one circle per country. In this view, the data are taken from the year 2015, with color indicating world region and the size of the circles representing Income per Person. Possible interactions in this system include changing the variables and settings, selecting countries either by clicking on the circles or on the country name list and watching the data over time through playback controls.

4.4 Markov Models

We propose that Markov chains can be used to model user interfaces and reveal potential biases in those interfaces. That is, we can model interface changes as a probabilistic sequence through a system's state space. We focus on the visual states that can be observed, leaving aside state changes that are only based on hidden internal representation changes.

A general *Markov model* is a statistical process that can be represented as a sequence of states and transition probabilities between those states (i.e., a state machine). Formally, let S_i for $i = 1, \ldots, n$ be a set of n possible states, and we define $P(S_i|S_j) = p_{ji}$ as the transition probability from state S_j to state S_i. A sequence of states may be thought of formally as $\{S_i, S_j, S_k, S_i, \ldots\}$, where a repeated state, like S_i represents re-visiting a state. All Markov models adhere to the *Markov property*,

Fig. 4.1 Gapminder Maps interface showing the initial state of the interface with 2015 data. See text for more details (Images from https://www.gapminder.org, CC-BY license)

which means transitions only depend on the current state (also called being "memoryless"). We represent this as the state of the system at time t being only a function of the state at time $t-1$, $P(S_t|\{S_1, S_2, \ldots, S_{t-1}\}) = P(S_t|S_{t-1})$.

The state machine model is the basis for other Markov processes. For example, a Markov chain is a path through a Markov model [15]. Hidden Markov models are Markov models that maximize the probability of observed chains when the underlying state space and probabilities are not known [1]. Markov models are "simple" in that they are amenable to many different kinds of analyses that yield useful information. Therefore, building a Markov model that faithfully reproduces system behaviors can lead to useful insights about expected behaviors under other circumstances.

The sequence of states in a Markov chain can represent a sequence of states the visual interface can go through. Those state changes maybe driven by direct user actions, streaming data updates or mixed-initiative analysis as it makes recommendations. The complete set of states in the Markov model is comprised of the union of all valid chains. This concept is illustrated in the Gapminder Bubbles visualization in Fig. 4.2. The three screenshots show a progression of states in the system. Figure 4.2(top) shows the initial view of the data when the year 2015 is selected. Figure 4.2(middle) shows the interface after the country India is selected by clicking on the India circle. Figure 4.2(bottom) shows the interface after the circle for Switzerland has been hovered over with the mouse. We note that a display changes can result from two types of changes. The first is a change in content/data produced by replaying the data over time with the playback controls. The second is a change in the layout or design parameters, resulting from a reconfiguration of the visualization through the right-side panel. To supply the transition probabilities, and thereby complete the Markov model, we assume that possible states of the interface are states in the Markov model and transition probabilities are derived from the screen presence of interface elements. A user session is a Markov chain, drawn from the probability space defined by the model. Analyzing the Markov model state machine provides insight into possible and probable user session patterns.

We can gain insight about potential system biases, pre-existing and technical, by examining the structure of connections between and understanding the relative likelihoods of interface states. For example, some states may not be reachable without a specific sequence of user actions, making them less likely to occur. Other states may have likelihoods that change over time because of certain design or algorithm choices. Still, others may be dependent on the default settings (the initial conditions) of the system. Modeling the user interface independent of actual user actions provides a basis for comparing interfaces to each other. Additionally, examining user interface actions in light of interface bias can tell you if observed biases came from the tool or from the operator. It allows us to distinguish the potential technical and pre-existing biases from the emergent biases in interactive visual analytic systems.

4.5   Interface Models

There are at least two conditions to interface modeling: with and without data loaded. With a dataset loaded, we propose to construct the Markov model with three key features: (1) each link is a possible action; (2) each node is an interface state that results from an action; and (3) links are weighted proportional to the target area on the screen. The above procedure captures the essential idea, but it probably needs to be tempered in some cases. Figure 4.3 illustrates some of the network shapes that result from applying this process by hand to parts of the Gapminder "Bubbles" interface. Linear dependencies are evident, showing that moving large distances in time incurs many step costs, biasing the user to make comparisons in near neighborhoods.

Applying the same procedure to the Gapminder "Map" interface (Fig. 4.1) yields similar patterns BUT with different weights. For example, in the "Bubbles" interface it is possible to directly select the Switzerland bubble, as in Fig. 4.2(top). However, in the map view, Switzerland is completely occluded by neighboring data. A data-dependent analysis of the interface would directly reveal this bias against such data points by examining the weights derived from screen space. Similarly, data-dependent analysis could reveal if the bias *toward* particular data points is proportional to bias in the dataset.

We have only done a partial analysis of the Gapminder interfaces, but we expect similar patterns to be components of full-application analysis. Just observing structural patterns, these patterns can illustrate potential biases. For example, isolated groupings show areas that may be difficult to move between - a bias for staying with the current representation. Moving to more algorithmic analysis, it would be possible to identify unreachable and difficult-to-access data.

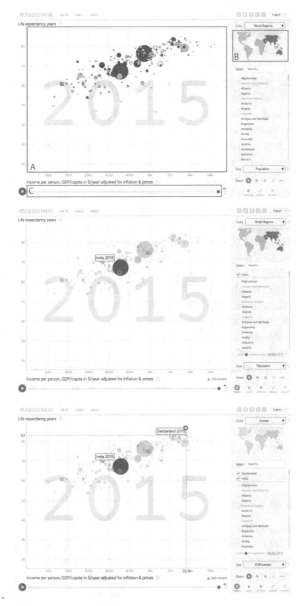

Fig. 4.2 Series of images of a Gapminder "Bubbles" view in sequential states: (top) initial 2015 data, (middle) select India, (bottom) hover Switzerland. Images from gapminder.org, CC-BY license. In the top image, (A) indicates the map plot window, (B) are the interface settings controls, and (C) is the playback controls to show animations over time. The data is shown with Income per Person on the *x*-axis and Life Expectancy (in years) on the *y*-axis. The size of the circles represent population, and the circles are colored by region

Modeling with weights in proportion to a target's area is at least partially justified by the Shannon entropy interpretation of Fitts' Law [8]. In brief, if a longer sequence of actions (or a sequence of more unlikely actions) is required to reach a state, that state is less likely to be encountered by chance. A sequence of user actions can be viewed as a string that encodes the address of an interface state. In terms of information, if bits of information must be supplied to "address" a state, the likelihood of an error increases. If there are more redundant paths, it is analogous to encoding redundancy and the state is more likely.

The Markov chain conceptualization for data-dependent biases derives the transition probabilities, p_{ji}, from this weighting schema. We are capturing biases where the transition probabilities shape Markov chains to end up in a particular part of the state space or make some transitions more likely than others. With data in the system, we are measuring some of the technical biases. The data representations reflect the results of the underlying encoding/embedding schemes and choice of machine learning or analytic algorithms. These technical choices can bias the data available in the system. Pre-existing biases may come into play if the system is applied to data types for which it was not designed, because the norms and practices will not properly apply. This would occur, for example, if numerical techniques are applied ineffectively to encode text data. But predominantly, data-dependent Markov chains capture technical system biases.

This preliminary analysis makes it evident that the basic procedure naïvely applied yields a combinatorial explosion of states. For example, sequential data selection is done when picking specific countries in the Gapminder "Bubbles" chart. A full model is a lattice of all possible combinations of selections (A, B, C, A&B, A&B&C, A&C, B&C, etc.). For all but trivial examples, this is likely to be computationally intractable. Tempering full data dependence is probably necessary and is the focus of the next section. In truth, a mixture of data-dependent and data-independent modeling is likely to yield the best *tractable* models. Some of the simplifications used in Dabek and Caban [4] reduce the impact of redundant combinations may also have analogous simplifications for this a priori modeling.

Data Independent Modeling

Interesting patterns in the interface may be revealed by ignoring details of the data presentation. In the data-independent scenario, the resulting model is simplified but necessarily more abstract. It is constructed in the same way as the data-dependent bias case but with two simplifications. First, all interactions that directly involve the data are collapsed into a single link by type. For example, instead of a selection-related link for each data point, there is a single data-selection link. This necessarily implies that data-related states are also compressed together. The general transformation is shown in the difference between the top and bottom row in the left column of Fig. 4.3. Second, because we are no longer considering the data representation, we can no longer use screen-space to weight the links. Instead, we propose to make all links that leave a node equally likely. This is termed a *regular* Markov chain, with the transition probability matrix $P = \left[\frac{1}{n}\right]$. This initial assumption provides a baseline against which we can study a system.

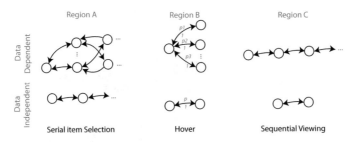

Fig. 4.3 Markov model structures from Gapminder "Bubbles" regions noted in Fig. 4.2(top). The difference between the data-dependent and data-independent cases is evident in the difference of complexity between the rows

Data-independent Markov chains have transition probabilities that are regular or are shaped by the initial conditions of the system. If the transition probabilities are dependent on initial conditions, we are capturing a pre-existing bias in the system. That is, the assumptions made by the designer as to default settings produced a bias toward data availability that changed when those default settings were adjusted to some alternative initial configuration. Additional pre-existing biases are captured in the overall design elements in the display or choices of representation implemented, because all reflect some methodological attitude or cultural norm for that system. Technical biases can also be revealed if the data-independent display incorporates structures output from some internal algorithm, or the structure reflects technology choices on which the system is implemented. But we argue that data-independent Markov chains serve to capture pre-existing system biases.

Modeling an interface with a specific dataset represented is likely to be more directly actionable than the data-independent model. However, the models are likely to be large relative to the data-independent case because many common interface patterns are combinatoric in the elements of the dataset. Working with the data-independent model has the effect of reducing the size the model significantly, but it makes the results more abstract and thus more difficult to interpret.

4.6 Application: Gapminder Analysis

We prototyped the bias measurement procedure on the Gapminder world map visualization. The target application is show in Fig. 4.1. The data-based components were recreated using Gapminder's demographic data [12] and geographic centroids [11]. Countries are represented by circles, the areas of which correspond to the income variable. Figure 4.4 shows the basic result of this abstraction.

Estimating bias according to the procedure outlined in Sect. 4.5 requires measuring the proportion of pixels allocated to various interactive elements. This can be accomplished by assigning each interactive element a unique color and counting how

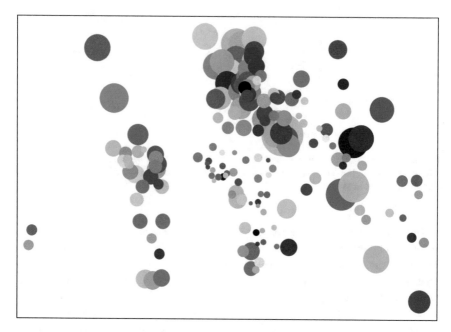

Fig. 4.4 Re-representation of Gapminder Maps with random colors for measuring bias. Circle size represents Income, and position of each circle is the same as in the Maps view in Fig. 4.1

many pixels in the end image contain each color. To properly measure the interface bias, an image must be measured for the interactive state of each interface element. In the Gapminder Map, the main map interaction is selecting countries. When a country is selected its label is rendered top-most and can be used for selection in the same way that the circles can be. Therefore, the measurement process creates a separate image for each interactive state. In this case, each image corresponds to selecting a different country and includes a label box for the selected country. The label box is filled by the same color as the country because, in the Gapminder map, country labels behave as selection targets in the same way the country's circle does. The underlying map is not directly interactive and is thus omitted from Fig. 4.4. The various controls on the periphery are also omitted from this analysis.

With an image similar to that in Fig. 4.4 generated for each country, the number of pixels allocated for each country can be counted directly. Because the background color is the most common color (comprising 88% of the image), it was omitted from this model. However, in modeling other interfaces it may be valuable to include. Each image corresponds to a state in the Markov model and the percent of pixels for each country corresponds to the transition probabilities.

With the Markov model defined, analysis can proceed. There are two basic measures: the baseline probability and the stable distribution. The baseline probability is the average probability across all possible transitions. Shown in Fig. 4.5 in comparison to the population distribution, the distributions are distinctly different.

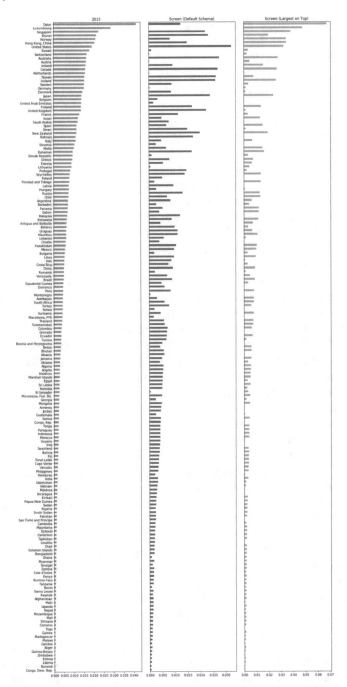

Fig. 4.5 Screen space and proportional incomes compared (ordered by screen space). If screen space were allocated proportional to income, the blue and green series of bars would both monotonically decrease. Because they do not, there is disproportionate representation

Treating screen and data proportion as a sorted list, the relationship between the two can be measured with Spearman's rank correlation (i.e., Spearman's ρ). ρ ranges from -1 to 1, corresponding to inversely ordered to identically ordered. A value of 0 indicates that the orders are unrelated, and is the null hypothesis. With an ordering of countries based on screen proportion and another based on data proportion, $\rho = -0.02$, with $p = 0.75$. Relative to the common type I error rate $\alpha = 0.05$, this result indicates that the orders cannot be distinguished from random. We must conclude that any bias in the visualization is not related the distribution in the source income data.

The other basic measure is to look at the stable distribution, essentially modeling what random walks across the interface would produce. The data-dependent case, where the actual data values are used to scale the circles, is shown in Fig. 4.6. It is clear that there is a significant bias towards specific countries, but that bias is not matched by the per-capita income of those countries. In fact, several highly populated countries are at the bottom of the distribution (Luxembourg, Switzerland, Austria, Netherlands, Germany, Montenegro, El Salvador), but all are in regions of the world with many political boundaries close together such as Europe and Central America. In contrast, the top of the distribution (United Kingdom, Brunei, Japan, Iceland, Taiwan, Canada, Australia, United States) is made of geographically isolated countries, even though they do not have the highest per-capita incomes.

Interpreting these results requires knowing what the *desired* outcome is. The argument for approximately equal distributions is that each item is equally selectable. Our analysis indicates that the Gapminder interface essentially supports this type of analysis when used interactively, but only when used interactively. In contrast, if a bias that follows the data distribution is desired (that the answers should "pop out") this layout fails both interactively (where distributions are too even) and statically (where the image does not allocate pixels proportional to the source data).

The data-independent analysis reveals limits about the interface regardless of exact data values. In this analysis, images were generated where each country was given the same value. Exploring different cases involved using different assigned values. To provide an even distribution statically or in the stable distribution at common screen (100 dpi) or print resolutions (300 dpi), the circle for each country would need to be smaller than a single pixel. This is impractical, and thus we conclude that the map layout provided is incapable of providing an even bias.

Our re-implementation of the Gapminder interface is not perfect. There are three main differences. First, Gapminder's actual interface uses an area-preserving geographic projection (or a compromise project that includes area-preservation as a partial criterion). For simplicity, we used an equi-rectangular projection. This does not affect the procedural validity, but it likely influences the exact weights in the Markov model as overlapping regions may shift around. It is likely that our analysis reports less bias than a matching projection because much of the bias is found in Europe, which is more compressed in most area-preserving projections than in equi-rectangular. Second, We have omitted the controls surrounding the main map and the background map itself. The background map was omitted because it is essentially non-interactive. Other controls were omitted for simplicity of analysis.

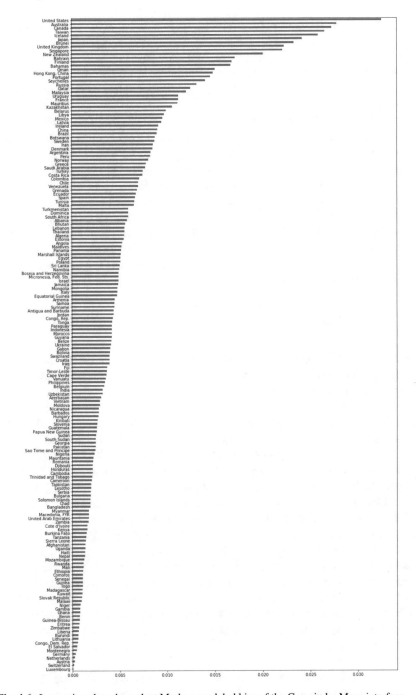

Fig. 4.6 Interactive, data-dependent Markov-modeled bias of the Gapminder Maps interface

Finally, the scaling factor used for circles approximates that of the Gapminder interface, but is not a perfect match.

Our analysis can be used to directly explore alternative implementation decisions. For example, to faithfully reproduce the Gapminder Maps interactive interface elements, country circles are rendered such that smaller values lay on top of larger values. This makes it more likely that small countries will be selected than their proportion of the data would indicate. Using our analysis techniques, we can also measure what the bias *would be* if countries were rendered in other orders. If large income countries were rendered on top, the interactive case appears more like the background data distribution (see Fig. 4.5, right column), but not sufficiently to be statistically significant (Spearman's $\rho = 0.06$, $p = 0.38$). This indicates that the bias is dominated by something other than rendering order.

Rendering the order by other data values would also provide other bias profiles, some of which may be useful for specific contexts (e.g., conditioning income maps by population may bias the interface towards discovering patterns in poverty). This approach provides opportunities to explore interface decisions and how they may be made in context-specific ways.

4.7 Discussion

Classic Markov modeling is a "memory free" technique. It only takes the current state into consideration when making a transition. However, data exploration necessarily includes human memory [16]. Modeling multi-step memory with static Markov models is cumbersome at best (and practically impossible in combinatoric cases). However, compressing combinatoric cases into abstract chains (as discussed earlier) can be seen as a simple memory model. A similar compression technique might be used to model a simple form of memory. An alternative to combinatoric compression of states would be to use a model that includes memory in a structured way. Dynamic Markov, Push-down automata, and RAM-based automata (with limited RAM) are also viable options. Each has a finite state space and a well-developed field of analysis.

Our proposed weighting scheme is simple, and may not be sufficient to illuminate some bias patterns. There are some interesting challenges. For example, in the data-dependent construction, the size-based weighting is derived from Fitts' Law. However, Fitts' Law does not account for convention or attention. Therefore, some interface elements may be relatively large by convention but the probability that they will be interacted with is not proportional to their size. For example, menu bars have a size and position dictated by the interface guidelines of the platform, and that may be significantly larger than the representation of a single data point. Capturing such differences in the interaction probabilities requires reaching beyond Fitts' Law for transition probabilities.

In the data-dependent Markov modeling, *only* the screen real-estate is used to model direct data interactions. Logical extensions include using visual similarity (along many retinal dimensions) to up-weight or down-weight items. This could be extended further with a dynamic Markov model, so weights change based on what states have been visited in earlier interactions. Proper dynamic weighting requires knowledge of the task as well as the visual representation. It makes sense to up-weight similar things when the retinal variables correspond to the desired task but to (possibly) down-weight similar items when the retinal variable does not have a bearing on the task. Also, exploration versus verification probably has different interaction patterns. Such modeling may be achieved using a Markov Decision Process. In addition to a transition probability, the model is extended with a payoff matrix and a "discount" factor. Payoffs are provided when a specific transition is taken. The discount factor determines whether immediate payoffs or future expected payoffs are prioritized. Decisions are still based on the information observable in the current state, but the probability of a transition is made a factor of the base probability, the payoff, the expected future payoff and the discount factor. Payoff and discount factors can be adjusted to model different goal-directed behaviors. Similar dynamic re-weighting is done in Dabek and Caban [4], captured in their "ideology" factors.

Analytic provenance models suggest another approach to Markov modeling. In particular, if a provenance tracking system records information about the state of the interface, we could use a hidden Markov model to derive the Markov chain of the original interface state space [7]. This might be helpful in cases where we have incomplete information about the structure or state space of an interface. This inference process could leverage existing graph modeling systems for analytic provenance, as in GraphTrail [5], to interpret the hidden model states. This approach bears some similarity to Jankun-Kelly's [14] P-set Model of visualization exploration. He defines two key concepts. A P-set is a set of parameters that define a visualization system, and visualization transformation is an operation on the P-set that creates a particular visualization view. Each set of parameter values (P-set) defines a state space with weighted connections (transformations) between the states. The difference between our Markov chain approach is that our links between the states quantify the probability of moving between states, rather than defining the parameter transformations themselves. An interesting direction for future work is to relate the transformations to transition probabilities between parameter states to capture emergent bias.

4.8 Conclusion

We note that methods for measuring information content in a visual analytic system remain an open challenge for the field [19]. Such measures are important for the overall evaluation of systems, particularly for calibrating our expectations for how much information users may be able to extract from a system. We propose that measurement of information availability and the interface biases that may shape that information availability should be modeled in systems before they are put into human-in-the-loop evaluations. Markov models, as proposed herein, provide a promising direction for conceptualizing the state space of a visual analytic system and understanding system-level biases through the transition probabilities over the state space.

Acknowledgements This research was sponsored by the Analysis in Motion Initiative at the Pacific Northwest National Laboratory. The views and conclusions contained in this document are those of the authors and should not be interpreted as representing the official policies, either expressed or implied, of the U.S. Government.

References

1. Ballard DH (1997) An introduction to national computation. MIT Press, Cambridge, MA
2. Cook K, Cramer N, Israel D, Wolverton M, Bruce J, Burtner R, Endert A (2015) Mixed-initiative visual analytics using task-driven recommendations. In: 2015 IEEE conference on visual analytics science and technology (VAST). IEEE, New York, pp 9–16
3. Cook KA, Thomas JJ (2005) Illuminating the path: the research and development agenda for visual analytics. IEEE Computer Society, Los Alamitos, CA
4. Dabek F, Caban JJ (2017) A grammar-based approach for modeling user interactions and generating suggestions during the data exploration process. IEEE Trans Visual Comput Graphics 23(1):41–50
5. Dunne C, Henry Riche N, Lee B, Metoyer R, Robertson G (2012) Graphtrail: analyzing large multivariate, heterogeneous networks while supporting exploration history. In: Proceedings of the SIGCHI conference on human factors in computing systems. ACM, New York, pp 1663–1672
6. Endert A, Fiaux P, North C (2012) Semantic interaction for sensemaking: inferring analytical reasoning for model steering. IEEE Trans Visual Comput Graphics 18(12):2879–2888
7. Endert A, Ribarsky W, Turkay C, Wong B, Nabney I, Blanco ID, Rossi F (2017) The state of the art in integrating machine learning into visual analytics. Comput Graphics Forum. https://doi.org/10.1111/cgf.13092
8. Fitts PM (1954) The information capacity of the human motor system in controlling the amplitude of movement. J Exp Psychol 47(6):381–391
9. Friedman B (1996) Value-sensitive design. Interactions 3(6):16–23
10. Friedman B, Nissenbaum H (1996) Bias in computer systems. ACM Trans n Inf Syst (TOIS) 14(3):330–347
11. Gapminder Foundation. Gapminder.org: Geography. https://www.gapminder.org/data/geo/ (2018). Accessed 01 Apr 2018
12. Gapminder Foundation. Gapminder.org: List of indicators in gapminder world. https://www.gapminder.org/data/ (2018). Accessed 01 Apr 2018
13. Gotz D, Zhou MX (2009) Characterizing users' visual analytic activity for insight provenance. Inf Visual 8(1):42–55
14. Jankun-Kelly T (2008) Using visualization process graphs to improve visualization exploration. In: International provenance and annotation workshop. Springer, Berlin, pp 78–91
15. Kemeny JG, Snell JL (1960) Finite Markov chains, vol 356. van Nostrand, Princeton, NJ
16. Patterson RE, Blaha LM, Grinstein GG, Liggett KK, Kaveney DE, Sheldon KC, Havig PR, Moore JA (2014) A human cognition framework for information visualization. Comput Graphics 42:42–58
17. Wall E, Blaha L, Franklin L, Endert A (2017) Warning, bias may occur: a proposed approach to detecting cognitive bias in interactive visual analytics. In: IEEE visual analytics science and technology (VAST). IEEE, New York
18. Wall E, Blaha L, Paul CL, Cook K, Endert A (2017) Four perspectives on human bias in visual analytics. In: Ellis G (ed) Cognitive biases in visualizations, Chap. 3. Springer, Berlin
19. Ward MO, Grinstein G, Keim D (2010) Interactive data visualization: foundations, techniques, and applications. CRC Press, Natick, MA
20. Wattenberg M (1999) Visualizing the stock market. In: CHI'99 extended abstracts on human factors in computing systems. ACM, New York, pp 188–189
21. Xu K, Attfield S, Jankun-Kelly T, Wheat A, Nguyen PH, Selvaraj N (2015) Analytic provenance for sensemaking: a research agenda. IEEE Comput Graphics Appl 35(3):56–64

Part II
Cognitive Biases in Action

Chapter 5
Methods for Discovering Cognitive Biases in a Visual Analytics Environment

Michael A. Bedek, Alexander Nussbaumer, Luca Huszar
and Dietrich Albert

5.1 Introduction

The development and application of new knowledge and information technologies have enormous influence on the way people live, work and learn. In the law enforcement sector, analysts are constantly required to understand and make sense of huge amounts of often unstructured data. Sense-making in this context means that analysts have to find and interpret relevant facts by actively constructing a meaningful and functional representation of some aspects of the "whole picture". Visual Analytics (VA) possesses the potential to support the analyst's reasoning and sense-making processes.

This is the point where the European project VALCRI[1] comes into play. Addressing the challenges of today's law enforcement agencies, the main aim of this project is to support analysts in their reasoning and sense-making processes by providing appropriate data analytics tools, applying the methods of visual analytics. Thereby, one key focus of this project is concerned with human issues, such as, how to mitigate or avoid cognitive bias that might be caused by such automated systems, how sense-making occurs in this context, and how information and knowledge should be structured to support the human reasoning process.

---

[1] http://www.valcri.org/.

M. A. Bedek (✉) · A. Nussbaumer · L. Huszar · D. Albert
Institute of Interactive Systems and Data Science,
Graz University of Technology, Graz, Austria
e-mail: michael.bedek@tugraz.at

A. Nussbaumer
e-mail: alexander.nussbaumer@tugraz.at

L. Huszar
e-mail: luca.huszar@tugraz.at

D. Albert
e-mail: dietrich.albert@tugraz.at

© Springer Nature Switzerland AG 2018
G. Ellis (ed.), *Cognitive Biases in Visualizations*,
https://doi.org/10.1007/978-3-319-95831-6_5

Fig. 5.1 This figure shows the time, location, and bar chart tool of the VALCRI platform

In the course of this project, a visual analytics platform has been created that addresses the functional and thinking requirements of analysts [23, 24]. This platform consists of more than fifteen synchronized tools. Five of them are described in the following and three of them are depicted in Fig. 5.1.

- The *Search tool* allows to search for specific crime incidents or to filter them on geographical area, time frames and crime types (e.g. burglary). The result is made accessible through the various tools of the VALCRI platform.
- The *Time tool* shows a line chart that indicates the number of crime incidents. The time frame can be changed interactively, in order to get either a more detailed view or an overview of the data. Similar to the Time tool, a Statistical process control tool (SPC-tool) shows standard deviations of the number of recorded crime incidents in this time frame. This allows the user to quickly spot statistical outliers which may indicate that something unusual happened.
- The *Location tool* depicts crime incidents on an interactive map. Crime incidents are represented as single dots or as rectangles, if a larger set of crimes are available in that area (more than 200). In such cases, the size of filled-out rectangles within a particular area indicates the number of crime incidents - the highest is completely filled and other areas are relative to this. The map can be interactively zoomed in and out, which changes automatically the visual representation and synchronizes the other tools with the updated dataset selection.
- The *Bar Chart tool* shows the number of crimes according to a classification scheme. Discrimination factors include crime types, districts and resolving state. According to such discriminators, the numbers of crimes are shown on a bar chart sorted by the number of crime incidents. Clicking on a particular bar limits the dataset and synchronizes the other tools accordingly.
- *The List tool* presents a list of the currently selected crimes including their meta-data. Details of the crime are shown including the involved subjects, the location, time information and full description.

Even if the support for sense-making with VA technologies is helpful and valuable, there is still a well-known problem of systematic errors, so-called cognitive biases, that might hinder analysts to draw sound conclusions. Cognitive biases occur when imperfect knowledge, uncertainty, complexity and time constraints prohibit people from making optimal decisions. In such situations, peoples often apply heuristics, which can be thought of as "rules of thumb" when making decisions or when evaluating the value, importance and meaning of information. These heuristics are useful in many cases, however, they can lead to severe and systematic errors in judgments and decisions [15, 21]. In the context of law enforcement analysis, these "systematic errors" or cognitive biases can occur in every phase of the decision making and reasoning process, such as discounting, misinterpreting, ignoring, rejection or overlooking pieces of information.

A large number of cognitive biases have been suggested and described in the literature. However, in the course of the VALCRI project and related requirements analysis, a set of eight cognitive biases has been selected, based on their significance for the daily routines of analysts [13]. These cognitive biases are listed in Table 5.1.

This chapter focuses on the question of how to ensure that a VA-platform mitigates cognitive biases from different perspectives: A (i) theory-driven, (ii) empirical and (iii) a data-driven perspective.

On the one hand, mitigating cognitive biases means reducing the probabilities that cognitive biases occur, or on the other hand, if they can not be avoided, to reduce their negative effects on the decisions and judgments. A prerequisite for answering this question empirically, for example in the course of experimental summative evaluations, is the measurement if and to what extent a cognitive bias occurs. Operationalization refers to the process and outcome of making non-directly observable constructs measurable. This would enable cognitive biases to be measured whilst a user interacts with a VA environment.

Table 5.1 Relevant cognitive biases in the VALCRI project

Cognitive bias	Description
Confirmation bias	Where pieces of information that support the initial expectation are disproportionally considered and selected [17]
Anchoring	Which is the tendency to rely too heavily upon or to "anchor" on a past reference or on one trait or piece of information when making decisions [16]
Clustering illusion	Which is a tendency to see patterns where no patterns exist, e.g. interpreting patterns or trends in random distributions [12]
Framing effect	Which is the tendency to draw different conclusions from the same information, depending on how that information is presented [22]
Availability bias	Where likelihood-estimations of something to happen is by the ease with which instances of occurrences can be brought to mind [21, p. 1127]
Base rate fallacy	Which is the tendency to base judgment on specifics, ignoring general statistical information [11]
Selective perception	Occurs when people pay particular attention to some parts of their environment to the point where it distorts the reality of the situation [5]
Group-think	Is a deterioration of mental efficiency, reality testing and moral judgment resulting from group pressure [14]

In the following section, we address some theory-driven approaches. Theory-driven refers to the fact that solely domain experts, in this case, experts in the field of cognitive science or cognitive biases, address the question of how to avoid, mitigate or operationalize cognitive biases. In the first subsection, some examples for a-priori design principles are given - for example how visualizations should be designed or how data should be represented. It is followed by a subsection on how to systematically analyze the tools which constitute a VA platform and a subsection which describes how to measure cognitive biases "on the fly", i.e. by identifying actions and interactions with the platform. The consecutive section deals with empirical approaches, such as behavioral observations of analysts and operationalizations of cognitive biases that enable us to carry out experimental studies. We call these approaches empirical, because end-users, i.e. analysts, are required and their data, responses and evaluations are used for data analysis. Finally, the data-driven approach refers to statistical and data-mining methods that aim to identify patterns of a user's interactions with the visual analytics platform that correlates with the presence or absence of cognitive biases.

5.2 Theory-Driven Approaches

This section describes three methods for cognitive bias detection and mitigation that are based on theoretical considerations and a literature review.

5.2.1 Design Recommendations

In the ideal case, visualizations are designed in a way that they do not induce cognitive biases at all. For several reasons, this ideal case is hard to achieve. Visualizations are made to serve a specific purpose, for example, to give an overview or to summarize data which could be only be described in confusing tables or exhausting texts. Representations are less detailed, less complex or less manifold than the part of the reality it aims to represent. Visualizations usually present a subset of a particular set of data; the more prototypical this subset, the easier it is for its recipients to generalize the whole dataset. The selection of subset and the way it is displayed, structured and visualized is the outcome of the human decision process of the visualization designer. However, human decision processes are vulnerable to cognitive biases. Nevertheless, a small set of a-priori design principles on how to make good visualizations can help. At least, there is a small set of recommendations on how to avoid some notable cognitive biases in visualizations, for example through the graphical layout of competing information [4] or through multiple views of the same information [13].

In the following, a simple example demonstrates how the above-mentioned selection process, as well as design decisions on how to display these pieces of information, might have an effect on recipients. One particular cognitive bias which has an impact on the selection process is called *Selective Perception* and a particular cognitive bias which has an impact on how a certain visualization is interpreted is the so-called *Framing Effect*. *Selective Perception* refers to the effect that only a small part of the reality is represented and in the focus of one's attention, a small part that is usually not representative of the whole. The *Framing Effect* is the tendency to draw different conclusions from the same information, depending on how that information is presented [22]. The data in the following chart (Fig. 5.2) is from the 2016 Annual report of the Police crime statistics of the Austrian Ministry of the Interior [3]. The data represent the overall numbers of recorded complaints. It demonstrates an example of the *Framing Effect*. In these two charts the same information is depicted with different aspect ratios. The chart on the left side uses an aspect ratio of 3:5, while the chart on the right side uses an aspect ratio of 4:3. The increase of complaints and records from the year 2015 to 2016 looks more dramatic in the left chart than on the right-hand one. Therefore, the American Psychological Association [1] recommends using a 4:3 aspect ratio for all histograms and bar graphs. The range of scales can also have a large effect. For example, when comparing Figs. 5.2 with 5.3 it becomes obvious that the increase from 2015 to 2016 becomes even less dramatic, if the ordinate starts at 0. The APA suggests to either start all ordinates at 0 or to clearly highlight it otherwise.

Figure 5.3 also indicates that the impression of trends is dependent on the time frame, which is also an example for *Clustering Illusion*. The chart at the right-hand side of Fig. 5.2 shows that the numbers are actually decreasing when comparing both halves of the 10-year period.

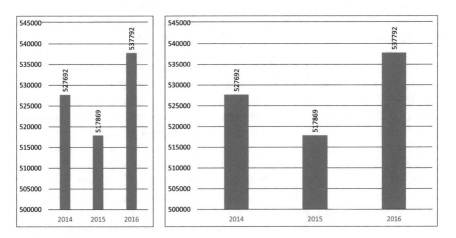

Fig. 5.2 The data from 2014 to 2016 in 3:5 format (left) and in 4:3 format (right)

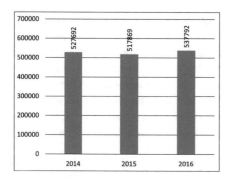 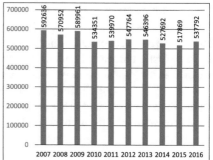

Fig. 5.3 The data from 2014 to 2016 with an ordinate starting from 0 (left) the data from 2007 to 2016 (right)

5.2.2 Systematic Tool Analysis

The systematic tool analysis aims to evaluate and improve VA environments and its tools with regard to their potential to avoid or mitigate cognitive biases. In a nutshell, this approach investigates each tool with respect to the various cognitive biases. In a first step, the tools of the platform are selected and briefly described, which includes the context in which they are used, their purpose, their input data and output format, etc. Then the tools are analyzed by domain experts such as cognitive psychologists or experts on cognitive biases. These experts have to evaluate if and, to what extent, the tools either mitigate or facilitate different cognitive bias. Ideally, such an analysis is done for each cognitive bias separately.

Such a systematic investigation leads to a matrix, with tools as rows and cognitive biases as columns. For each cell, the investigator describes to what extent the respective tool mitigates or facilitates that particular cognitive bias. For example, the *Map tool* (see Fig. 5.1) may lead to *Selective Perception*, if a specific area is heavily crowded with crime incidents, because this might attract the attention of the analysts away from other parts of the map. An example tool that has the potential to mitigate the *Confirmation Bias* is the *Time tool*, as it allows the user to change the time-frame and thus the amount of crime data displayed. This results in the presentation of different perspectives and contexts of crime data, which has the potential to avoid the *Confirmation Bias*. The outcome of this method provides an overview of the mitigation capabilities and dangers of cognitive biases of the whole platform.

In order to analyze the danger of cognitive biases and mitigation strategies of individual tools, we propose to follow the Delphi method [19]. Delphi is designed as a structured and systematic process to develop forecasting perspectives on future events by asking panel experts. Typically, this method is performed in two or more iterative rounds, whereby in each round, every expert evaluates the current state and provides additional input, which leads to an adapted and improved next version.

5.2.3   Process-Oriented Operationalization

The aim of the following approach is to identify and describe the users actions and interactions with the tools of the VALCRI platform, in order to measure their tendencies towards cognitive biases. The effect of a particular cognitive bias can be predicted in certain well-defined decision tasks. However, in the case of an interactive VA platform, there is a wide range of potential behavioral manifestations that makes it impossible to describe all the actions and interactions which occur when a biased behavior takes place. The design recommendations and the systematic tool analysis described above can provide conductive insights that helps to identify the behavioral patterns related to cognitive biases. To demonstrate how this method works, we focus on the example of *Selective Perception* and briefly outline how it could occur by using the *Search*, *List* and *Location tool* of the VALCRI platform. As mentioned above, this cognitive bias is defined as being focused on a particular area of the information space, whilst ignoring other pieces of information.

To detect this particular cognitive bias, a similarity measurement can be computed between the keywords entered into the *Search tool*, i.e. between the documents and crime reports further examined via the *List tool* or between the parameters of the visualizations of the *Location tool*. A high similarity between the keywords, the selected documents and the visualization parameters over a longer period of time is considered as an indication that the user is focused on a particular area of the information space, i.e. the *Selective Perception*.

In the context of VA, it is important to distinguish between different kinds of searching modes, such as explorative, investigative, hypothesis-driven and question-driven searches. The validity of the operationalization of any cognitive bias can be improved when taking such contextual information into account. For example, in case of a hypothesis-driven search, an analyst who is engaged in a small area of the information space shouldn't be identified as being affected by *Selective Perception*, however, this does not mean that the user's behavior is not influenced by any other cognitive biases.

5.3   Empirical Approaches – Behavioral Observation and Outcome-Oriented Operationalization

This section presents two empirical methods for detecting cognitive biases.

5.3.1   Behavioral Observation

In the context of the VALCRI project, several behavioral observations have been carried out. In one study, nine experienced law-enforcement analysts worked on a task for around 2 h, separately from each other. While working on the task, they were asked to "think aloud" on their reasoning, ideas and conceptions. Their activities were video and audio recorded and the screen activity was captured. The participating analysts' task was to analyze a particular crime type in a city district over a given period of time and the main question for them was, should more patrols be sent to this city district. A qualitative interview was then carried out.

While working on the task, the participants were observed by at least one expert on cognitive biases who did not intervene during this exercise. The observer filled out a prepared form, indicating the time when a cognitive bias was observed, the tools that had been used by the analyst, and if necessary, further explanation on this observation in an open format. These observations were subsequently validated and enriched by two other experts who used the video and audio recordings.

On the one hand, the outcome of this exercise was a validation and enrichment of the systematic tool analysis described in Sect. 5.2.2, as well as the elaboration of new ideas for potential process-oriented indicators. On the other hand, compared to the purely theory-driven elaboration of the tool - cognitive bias matrix, the outcome of this exercise resulted in a mapping between sets of tools and cognitive biases. The reason for this is that for certain, often more complex, workflows and processes, the analysts used a combination of tools simultaneously.

An example would be the combination of the *Time tool*, the *SPC tool* and the *Location tool* when searching for "peaks in the noise", for a certain area and period of time. In many cases, the search for such peaks was focused on the maximum values and quite often, the analysts were not trying to falsify their initial hypothesis (e.g. by checking also for other periods of time or other city districts). This particular work process often resulted in vastly overlapping combinations of some cognitive biases: the *Confirmation Bias*, the *Framing Effect*, the *Base Rate Fallacy* and the *Clustering Illusion*, i.e. these cognitive biases occurred often in parallel.

5.3.2   Outcome-Oriented Operationalization

5.3.2.1   Confirmation Bias

Considering the large number of cognitive biases mentioned in the literature, only a few methods have been suggested for their objective measurement, such as a questionnaire or test. One example is the *Selective Exposure Paradigm* which has been proposed by Festinger [7] in the context of the cognitive dissonance theory, but later applied to elicit "confirmatory information search" [9]. Confirmatory information search is a main component of the *Confirmation Bias*. The *Selective Exposure Paradigm* is structured as follows: participants are confronted with a decision task and have to make an initial decision for one of two alternatives. Then the participants are exposed to various pieces of information that either confirm or disconfirm their initial decision. Half of the pieces of information are consistent with regard to the initial decision (i.e., the selected alternative) and half of them are not. In some cases of the *Selective Exposure Paradigm*, the pieces of information are short headline-like statements and the participants also have to indicate whether or not they would like to read further (more detailed) information on each statement [10]. Confirmatory information search is observed when a participant doesn't change their initial decision, even if overwhelmed by a large number of disconfirming pieces of information and if they are not interested in reading the detailed information.

Another aspect of the *Confirmation Bias* is *Confirmatory Information Evaluation* [8]. For each piece of information and statement, participants can be asked to what extend they consider this statement as important and credible. Importance and assumed credibility are usually highly correlated with each other. *Confirmatory Information Evaluation* can be observed if the importance and credibility evaluations for consistent statements (i.e. statements that are in favor of the initial decision) are higher than for statements that are in favor of the alternative.

The values for *Confirmatory Information Search* and *Confirmatory Information Evaluation* can be interpreted as an individuals' baseline-measurement of having a *Confirmation Bias* when evaluating the visualization system.

5.3.2.2   Clustering Illusion

The *Clustering Illusion* is defined as the tendency to see patterns where no patterns exist [12]. This tendency can be, for example, observed when people interpret patterns or trends in random distributions. A very similar cognitive bias is the Gambler's Fallacy, which refers to the belief that runs of one binary outcome will be balanced by the opposite outcome [2, p. 118]. In both cases, the cognitive fallacy is based on the belief that random events or data-points follow some rules, trends or patterns, which of course, they do not.

In the context of the VALCRI project, the following outcome-oriented operationalization of the *Clustering Illusion* has been applied: participants were confronted with a small dataset of 60 crime incidents and were asked to make a decision by means of the examples. They used certain tools of the VALCRI platform, in particular the *Location, Time* and *List tool*. The *Location tool* indicated the spatial distribution of crime incidents, the *Time tool* enabled to get insights on the temporal distribution of those crime incidents in different period of time and the *List tool* enabled them to look at some details of the incidents. The crime incidents had been randomly selected from a larger data-set and were located in two separate district of the suburban areas of the city of Birmingham.

In the main study, four examples were provided to the participants. For each example, the participants had ten minutes to inspect the data by using the above mentioned tools. Two examples were considered as random and the remaining two had been constructed in a way that there was a temporal increase for a period of six months and a local concentration within one of the city district. Another independent variable in the main study was the extent to which the participants could interact with the data. In half of the examples, the participants were allowed to interact with the tools and to change the parameters of the visualizations. In the other half, participants were asked to use only the *List tool* and to keep the other tools, i.e. the visualizations in the narrower sense, as prepared by the evaluators. In the interactive condition, it was possible to inspect the data from different perspectives and to principally falsify one's own impressions of patterns or trends.

After inspecting the data, the participants were asked (i) to evaluate if they would increase the police presence either in city district A or in city district B, (ii) to evaluate the certainty of their decision, (iii) to announce if their decision was based on the data or patterns and trends in the data, and if yes (iv) argue their decision. The idea was to measure an individual's tendency to see patterns where no patterns exist by the confidence ratings (ii) and the extent to which their decisions were based on data (iii) for the random-examples. These individual tendencies can be taken into account as the baseline when evaluating the visualization quality of the VALCRI system with regard to the *Clustering Illusion*.

5.4 Automatic Cognitive Bias Detection Approach

In this section, we briefly outline a method to automatically detect the cognitive biases based on user interaction patterns. Even if this method could be regarded as an approach that can be applied on any cognitive bias, we focus here on the *Confirmation Bias* and the *Clustering Illusion*. In addition to the automated bias detection method, it also outlines how a detected bias can be mitigated through feedback and prompts. This approach follows and extends the idea described by Nussbaumer et al. [18].

The starting point for the automatic cognitive bias detection is the operationalization as described in Sects. 5.2.3 and 5.3.2. They allow us to assess, in a controlled setting, whether a participant in such an experiment has these cognitive biases. Based on this method, we propose a data-driven approach to detect cognitive biases by taking into account interaction data of users (log data of user actions). If a cognitive bias is detected (indicated through a high probability for the occurrence of a bias), then a prompt or visual feedback is provided to the user (see Fig. 5.4).

The data-driven method is based on machine learning algorithms to automatically classify the users behavior in a visual learning environment based on the interactions with the tools of this environment. Participants in a study have to solve a criminal analysis task with the VAE. This task is embedded in a controlled experiment (e.g. *Selective Exposure Paradigm* described in Sect. 5.3.2.1 or the *Clustering Illusion* study described in Sect. 5.3.2.2) so that it can be assessed if their behavior is biased. Additionally, log data from their interaction with the VA tools are collected. From the experiment, it is known which interaction data is from biased and unbiased users and can subsequently create two groups. These two groups form the basis for further classification of interaction data from users that did not participate in a *Selective Exposure Paradigm*. In this way, when a user makes use of the VA tools, interaction data is collected and it can be determined if this interaction data is more similar to that of a biased or unbiased user. For clustering, several machine learning meth-

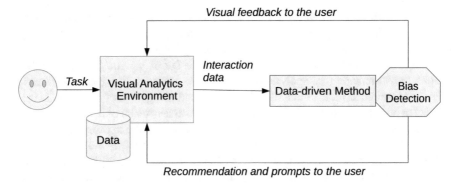

Fig. 5.4 This diagram depicts the overall approach to integrate automatic bias detection into a visual analytics environment

ods are available, such as the Support Vector Machine algorithm [20] or clustering algorithms [6].

The method described above, calculates probabilities for the occurrence of a cognitive bias. If such a probability is high, feedback could be provided to make the user aware that a cognitive bias might be involved in the thinking process. Such feedback can consist in visual clues that do not distract the user unduly, but nevertheless catches the users attention.

5.5 Conclusion and Outlook

Overall, this chapter aims at providing new methods and knowledge for discovering, measuring, and mitigating cognitive biases in the context of VA. Though a vast body of literature exists that deals with cognitive biases, most of it treats cognitive biases on a theoretical level. The work presented in this chapter includes several steps towards devising methods for measuring and mitigating cognitive biases.

Our elaborated methods extend the use of state-of-the-art of measuring cognitive biases on several dimensions. Firstly, a new procedure to measure the *Clustering Illusion* has been developed. The results are promising, but the applied methodology should be improved - further analysis of the log data should be carried out to determine whether or not it contains typical patterns of participants who are more influenced by the *Clustering Illusion*. Secondly, the method to measure cognitive biases through a classification of cognitive processes and assigning them in a structured observation constitutes a new approach in this field. This provides a basis for the operationalization of further cognitive biases. Thirdly, the data-driven approach outlines a method to detect cognitive biases based on user interactions with a VAE. All these methods outlines new directions on how cognitive biases can be measured, consisting of empirical studies, expert-driven behavioral observations and automatic observations through a logging system. In order to avoid detrimental effects of cognitive biases all together, new design recommendations have been elaborated. Though these design recommendations are based on existing ideas in literature, the innovation lies in in the translation of these ideas into the design of VA components. Furthermore, the systematic tool analysis provides a new approach to critically evaluate a VAE according to their potential inducements and mitigation of cognitive biases. This analysis allows for formative and summative assessments of a VAE.

Data visualization is a type of communication and just like in every communication process, the presented information could be misinterpreted by the receivers. The reason for this misunderstanding could be the presence of cognitive biases. In this chapter, we focused on a small set of cognitive biases, which could occur in a VAE. In the future, the design and evaluation of visualization techniques should be influenced by a combination of data-driven and theory-driven methods. The basic principles of these approaches could be easily transferred to different VAEs and applied on other cognitive biases. Another important aspect is the context in which the visualization is used. Ignoring the context, could lead to false classifications of biased and unbiased behavioral patterns.

The users require interactive interfaces and personalized visualization techniques. The appearance of emerging and innovative visualization techniques allows the user to interact in new way with datasets. Even if VA is a dynamic field of research, classical principles to detect and mitigate cognitive biases have been often disregarded. To design informative visualization with the least impact of cognitive biases, the cooperation of different fields of expertise is necessary.

Acknowledgements The research leading to the results reported here has received funding from the European Union Seventh Framework Programme through Project VALCRI, European Commission Grant Agreement N° FP7-IP-608142, awarded to B.L. William Wong, Middlesex University London and Partners.

References

1. APA (2009) Publication Manual of the American Psychological Association, 6th edn. American Psychological Association, Washington DC, USA
2. Barron G, Leider S (2010) The role of experience in the Gambler's Fallacy. J Behav Decis Making 23(1):117–129
3. BMI (2017) Sicherheit 2016 - Kriminalitaetsentwicklung fuer Oesterreich. Bundesministerium fuer Inneres. http://bundeskriminalamt.at/501/files/BroschuereSicherheit_2016.pdf
4. Cook M, Smallman H (2008) Human factors of the confirmation bias in intelligence analysis: decision support from graphical evidence landscapes. Hum Factors: J Hum Factors Ergon Soc 50(5):745–754
5. Dearborn DC, Simon HA (1958) Selective perception: a note on the departmental identification of executives. Sociometry 21(2):140–144
6. Everitt BS, Landau S, Leese M, Stahl D (2011) Cluster analysis. Wiley, New York
7. Festinger L (1957) A theory of cognitive dissonance. Stanford University Press, Stanford

8. Fischer P, Fischer J, Weisweiler S, Frey D (2010) Selective exposure to information: how different modes of decision making affect subsequent confirmatory information processing. Br J Soc Psychol 49(4):871–881
9. Fischer P, Greitemeyer T, Frey D (2008) Self-regulation and selective exposure: the impact of depleted self-regulation resources on confirmatory information processing. J Pers Soc Psychol 94(3):382–395
10. Fischer P, Schulz-Hardt S, Frey D (2008) Selective exposure and information quantity: how different information decision makers' preference for consistent and inconsistent information. J Pers Soc Psychol 94(2):231–244
11. Gigerenzer G (1994) Why the distinction between single-event probabilities and frequencies is important for psychology (and vice versa). Wiley, Chichester, pp 129–161
12. Gilovich TD (1991) How we know what isnt so: the fallibility of human reason in everyday life. The Free Press, New York
13. Hillemann EC, Nussbaumer A, Albert D (2015) The role of cognitive biases in criminal intelligence analysis and approaches for their mitigation. In: Brynielsson J, Yap MH (eds) Proceedings of the European intelligence and security informatics conference (EISIC 2015). IEEE, New York, USA, pp 125–128
14. Jones P, Roelofsma P (2000) The potential for social contextual and group bias in team decision-making: biases, conditions and psychological mechanisms. Ergonomics 43(8):1129–1152
15. Kahneman D (2011) Thinking, fast and slow. Farrar, Straus and Giroux, New York
16. Mussweiler T (2002) The malleability of anchoring effects. Exp Psychol 49(1):67–72
17. Nickerson RS (1998) Confirmation bias: a ubiquitous phenomenon in many guises. Rev Gen Psychol 2(2):175–220
18. Nussbaumer A, Verbert K, Hillemann EC, Bedek MA, Albert D (2016) A framework for cognitive bias detection and feedback in a visual analytics environment. In: Brynielsson J, Johansson F (eds) Proceedings of European intelligence and security informatics conference (EISIC 2016). IEEE, New York, pp 148–151
19. Okoli C, Pawlowski SD (2004) The Delphi method as a research tool: an example, design considerations and applications. Inf Manage 42(1):15–29
20. Steinwart I, Christmann A (2008) Support vector machines. Springer, New York. https://doi.org/10.1007/978-0-387-77242-4
21. Tversky A, Kahneman D (1974) Judgment under uncertainty: heuristics and biases. Science 185(4157):1124–1131
22. Tversky A, Kahneman D (1981) The framing of decisions and the psychology of choice. Science 211(4481):453–458
23. Wong BLW (2014) How analysts think (?): early observations. In: Proceedings of the IEEE joint intelligence and security informatics conference. IEEE, New York, pp 296–299
24. Wong BLW, Kodagoda N (2015) How analysts think: inference making strategies. In: Proceedings of the human factors and ergonomics society annual meeting, pp 269–273

Chapter 6
Experts' Familiarity Versus Optimality of Visualization Design: How Familiarity Affects Perceived and Objective Task Performance

Aritra Dasgupta

6.1 Introduction

Visualization techniques and systems are generally evaluated based on their perceptual effectiveness in supporting analytical tasks. Design principles and heuristics help guide the mapping across tasks, data types and visual representations of the data. However, visualizations designed and used by domain experts, are often in conflict with the established best practices. Examples include the use of the well-known rainbow color map, 3D-based encoding for non-spatial multidimensional data, spaghetti plots for showing temporal change in numerical data [1], use of many symbols for encoding categorical data [4], etc. Experts are often reluctant to use alternative methods to visualize their data, and in most of these cases, disagree about the negative effects of a design problem [4].

We attribute the factors causing experts' skepticism to the *familiarity heuristic*: experts use familiarity as a heuristic for subjectively preferring known methods over new ones and for being averse towards adopting a change in their existing visualization methods [16], although the familiar methods might be perceptually sub-optimal in performance.

In the cognitive science literature, the familiarity heuristic [2] is associated with the bias of availability [19] that suggests "*the likelihood of events is estimated based on how many examples of such events come to mind*". The more familiar a person is with the events, the easier it is to recall them and accordingly indicate a preference for them when faced with choices. For example, consumer behavior is guided by the familiarity heuristic, where people will tend to buy products of brands they are most familiar with [13].

Similarly, experts in different domains tend to use and adopt visualization methods and techniques that are conventional norms in their respective domains, despite their

A. Dasgupta (✉)
New Jersey Institute of Technology, New Jersey, USA
e-mail: aritra.dasgupta@njit.edu

© Springer Nature Switzerland AG 2018
G. Ellis (ed.), *Cognitive Biases in Visualizations*,
https://doi.org/10.1007/978-3-319-95831-6_6

Fig. 6.1 Familiarity of experts' versus optimality of visualization design. Recent studies have shown how familiarity affects usage and design patterns of visualizations by domain experts. Since familiar visualizations can often be in conflict with optimal choices, it is important to study how biases, due to familiarity, affect experts' trust, preference and objective task accuracy

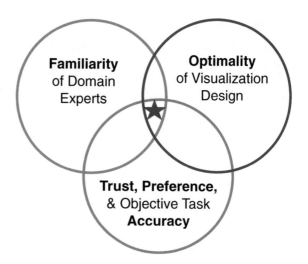

potential shortcomings. This is due to both a lack of awareness about the benefits of adhering to visualization best practices and a high degree of confidence in using the methods that experts are most familiar with.

In this chapter, we present a first analysis of the factors that are associated with familiarity related cognitive biases with respect to the three dimensions shown in Fig. 6.1. Namely, (i) experts' familiarity with visualization types and encodings, (ii) alternative optimal choices, and (iii) effects of familiar and optimal visual representations on experts' subjective perceptions (e.g. trust, preference) and objective task accuracy. We bring together knowledge gained from past studies to provide examples of manifestations of the familiarity heuristic in domain experts, provide examples of effects of the bias on expert judgment and discuss research questions that need to be addressed to help detect and mitigate the effects of the bias.

6.2 Manifestations of the Familiarity Heuristic

In this section, we borrow concepts from the cognitive science literature and discuss how biases associated with the familiarity heuristic manifest in experts' design and usage of visualization methods, techniques and systems. In the human-computer interaction literature, the term "intuitive" is often used interchangeably with "familiar". But, this work follows the recommendation of previous research [15] where it has been argued that "intuitive" can be misleading. Use of the term "familiar" helps us contextualize the use of visualizations across diverse domains, where different established practices may exist and thus the degree of familiarity with different visualizations may vary accordingly.

Perceived ease of use. The Technology Acceptance Model [8] prescribes that acceptance of computer-based techniques is largely dependent on their perceived ease of use, which is defined as *"the degree to which a person believes that using a system would be free of effort"*. One of the antecedents of perceived ease of use is self-efficacy, which qualifies how confident a person is in their own abilities to achieve the desired outcome. Self-efficacy is affected by the degree of familiarity with the task at hand [9, 17].

In the context of visualization usage, the perceived ease of use factor manifests in two cases: (i) where experts are more familiar with conventional, hypothesis-driven analysis methods, and (ii) where experts design visualizations using familiar tools that can have bad defaults or have inadequate support for analytical tasks.

As compared to other computer-based tools used for data analysis, like the use of scripting languages or Excel, visualizations are a relatively new way for many experts to interact with or present their data. Especially in domains where the use of static scripts facilitates hypothesis-driven analysis, experts often hesitate to adopt a data-driven approach using dynamic visualizations. This is mainly due to the low self-efficacy in switching contexts between understanding their data and formulating alternative hypotheses on the fly [6].

In many science domains, where experts design their own visualizations, the tools they use sometimes have bad defaults. But due to the high self-efficacy of experts in using those familiar tools and a lack of awareness of how the choice of visualization methods can affect tasks in practice, they prefer not to alter the defaults. In a study with climate scientists [4], we showed that such lack of awareness can lead experts to disagree with visualization researchers about the implications of poor design choices, especially those related to perceptual factors like clutter, color, etc. Subjective preference for the familiar, yet potentially less useful, visualizations is a natural connotation of the perceived ease of use factor.

Perceived accuracy and preference. Domain experts often believe that they will be more accurate in their analytics tasks using familiar ways of encoding the data. A popular example is the use of the rainbow color map in climate science. Despite the well-documented perceptual problems with the rainbow color map, scientists continue to use and prefer the rainbow color map. In a recent study [7] involving experienced climate scientists, it has been shown that the perceived accuracy levels using the rainbow color map (Fig. 6.2) is significantly greater than either of the choice that were less familiar, yet potentially more optimal, given the spatial data analysis tasks. This is one of the factors why scientists continue to have a high preference for the rainbow color map and many scientific data analysis tools still use it as their default.

Loss Aversion. Loss aversion [20] refers to the tendency of people to focus more on avoiding losses than on acquiring gains due to the perceived psychological impact of losses. This tendency manifests in the context of visualization design when a domain expert plays the role of a data producer or an analyst interchangeably. For example, climate scientists generate modeling data from simulation experiments for describing different phenomena in the earth and the atmosphere, and frequently

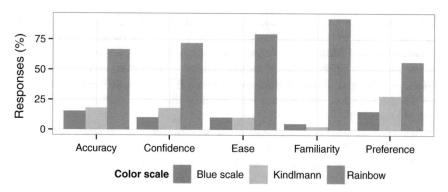

Fig. 6.2 Subjective rating of climate scientists about alternative color maps in a user study
The familiar rainbow color map (RBW) was clearly rated higher than the blue (BLU) or the Kindl-man (KIN) color scales based on perceived accuracy, confidence and ease of use [7]

use visualizations to communicate key messages about the model outcomes to stakeholders within and outside their community. In the course of our interactions with climate scientists [4], we found that even when the message could be conveyed by using abstractions or aggregations, they tended to focus more on avoiding loss of data in their visualizations, than on optimizing the visualizations for gaining insight from them.

This loss aversion tendency resulted in encoding information at a high level of detail, thus causing clutter in the visualization. In Fig. 6.3, we show two examples of this problem with a scatter plot with multiple symbols and a spaghetti plot with many overlapping lines. This tendency can be attributed to the confusion regarding the goal of a visualization: visualizations used for exploration or analysis are not optimal choices for communicating a message. Scientists, who create these visualizations, can easily find patterns in the data due to their high familiarity with the data, but others would not be able to spot the same pattern easily unless they are emphasized enough in the visualizations. These visualizations can, therefore, be suited to scientists' own analysis, but are ill-equipped to communicate a message to a broad audience, unfamiliar with the data or the problem domain.

Experts' trust: Similar to interpersonal relationships, in case of human-machine communication, familiarity breeds trust [11]. In a study [18] examining levels of system administrators' trust in familiar command-line interfaces as opposed to un-familiar graphical user interfaces (GUI), one of the participants remarked: *"Please, no more GUI. If people need a GUI, they aren't qualified to be doing whatever they are trying to do."* This quote is indicative and representative of the effect of the familiarity heuristic used by most of the participants: an overwhelming majority of them recorded a greater level of trust in command line interfaces although close to half the number of participants indicated greater levels of perceived ease with the GUI. This leads to a follow-on question - if a new analysis medium is able to better solve a problem than the more familiar ones, will experts trust the new medium?

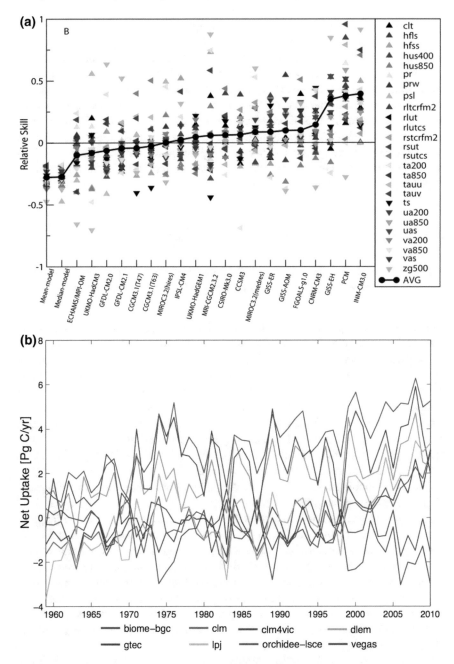

Fig. 6.3 Examples of familiar visualizations used for climate model comparison. The use of multiple symbols [10] in (**a**) and that of multiple overlapping lines in the spaghetti plot in (**b**), cause clutter, distract from the main message about similarity of model outputs and result from the experts aversion to loss

A recent study comparing the use of static scripts with that of a visualization-based system [6] addresses this question. Domain experts, despite their prior familiarity with static scripts, expressed comparable or greater levels of trust in a new visual analytic system. This was true especially in case of complex interpretation tasks where experts had to synthesize insights derived from multiple views of the data to confirm or refute their hypotheses. Similar to the study by Takayama et al., experts also expressed greater levels of perceived ease with the unfamiliar tool. The difference in the study set up here, as opposed to that study, was the fact that the unfamiliar tool was designed through a participatory design phase with senior researchers. This helped mitigate the potential concerns, due to lack of familiarity, of the bigger group of participants who had never seen the tool before.

6.3 Effects on Visualization Based Judgments

In this section, we describe how the familiarity heuristic manifests in domain experts' subjective and objective analytical judgments by reflecting on results from recent user studies.

Recurring design problems: In this study [4], we collected about 100 different visualization examples (e.g. maps, scatter plots and line charts) that are most frequently used in climate modeling for visually expressing similarity among multiple models. We then developed a classification scheme for describing the most common design problems (e.g. clutter, choice of visual variables and color map, etc.) and their consequences (e.g. misinterpretation, inefficiency, lack of expressiveness, etc.). As a next step, we discussed these problems with a group of climate scientists with a two-fold goal: (i) identify cases where experts and visualization researchers agree and disagree about the problem through interviews, and (ii) develop solutions to those problems and record their subjective feedback.

We found that in most of the cases, the majority of climate scientists disagreed about the existence and potential consequence of a design problem. Many of these cases involved serious consequences such as inaccurate judgment due to inappropriate use of a color map or due to the use of an inappropriate chart that did not adequately convey the intended message. Through our interviews, we saw a clear use of the familiarity heuristic, especially in the case of the most frequent design problems like the use of a rainbow color map or the use of multiple symbols on a scatter plot. This is reflected in the following comment where use of an alternative color map is perceived as a means to improve the aesthetics and not as a means to solve the task: "*I agree that the color map can be better but that would be a cosmetic change and wont affect the outcome.*"

We found that the effect of the familiarity heuristic could be mitigated in some cases when we collaboratively designed solutions for a subset of the familiar yet problematic visualizations. The solutions were designed keeping in mind the loss aversion tendency - the encoding could retain the fidelity of the data as much as possible, while at the same time, convey the main message about models that are

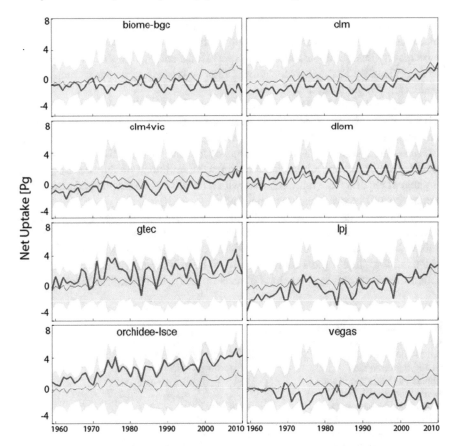

Fig. 6.4 **Modification of the spaghetti plot** (Fig. 6.1b) **into a small multiples of line charts** could overcome the familiarity barrier and was preferred by a group of climate scientists for visually communicating similarities among multiple model outputs

similar or different. For example, as a solution to the spaghetti plot (Fig. 6.3b), we designed a visualization with small multiples of line charts (Fig. 6.4), where each line chart represented a model and in each of them, one could directly compare the value for a particular model output with the mean and standard deviation of the sample. From experts' subjective feedback, we found that they were convinced that this would be an exemplary visualization that could potentially replace the spaghetti plot for comparing the temporal variation of multiple model outputs.

Discrepancy between subjective preferences and objective performance: We conducted a controlled experiment with a large group (47 participants) of climate modelers [5] to study the degree to which the familiarity heuristic affects objective task performance and also to analyze if there are discrepancies between subjective impressions like perceived confidence, accuracy, etc. and objective accuracy. We selected four visualizations, three familiar ones (heat map, bar chart and Taylor plot) and an unfamiliar one (slope plot), that were most suited to similarity and dissimilarity analysis tasks. The unfamiliar visualization was developed through a participatory design process with two experts for resolving the shortcomings of the familiar visualizations with respect to simultaneous comparison across many (>10) models and output variables.

We recorded prior levels of experts' familiarity with each of those visualizations and after the study, recorded their preferences and perceived levels of comfort, accuracy, etc. Besides an objective accuracy metric, we devised a *discrepancy* metric that measured the difference in rank orderings of the four visualizations based on their accuracy and based on their subjective ratings of familiarity, preference, etc. This let us gauge if experts were more accurate with a familiar visualization and also if their preference and accuracy rankings matched.

Overall, we found that perceptually motivated visualization design was a bigger driver for objective accuracy with and subjective preference for a particular visualization. In fact, in the case of a dissimilarity analysis task, the sub-optimal design of the familiar Taylor plots caused experts to be less accurate than when they used unfamiliar visualizations like slope plots (Fig. 6.5a), where explicit visual cues for

Fig. 6.5 Familiarity Versus Task Accuracy. We found that for the task of identifying dissimilar climate models, experts were more accurate with the relatively unfamiliar slope plots than the more familiar Taylor plots. In **a** we show the differences in objective performance accuracy. In **b** we show the discrepancy between rankings of visualizations derived from self-assigned familiarity scores and the rankings based on performance accuracy. We found statistically significant differences in both (**a**) and (**b**)

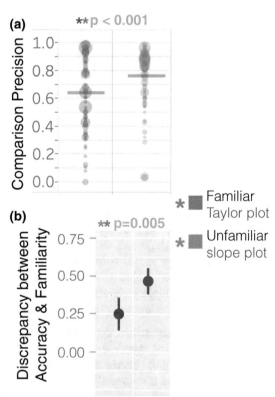

similarity and dissimilarity were encoded. Discrepancies could be observed between accuracy and familiarity rankings experts, across high and low experience groups, being more accurate with a less familiar visualization (Fig. 6.5b). Experts were also most accurate with their preferred visualization. The difference in preference levels for an unfamiliar visualization as opposed to a familiar one was less pronounced for participants with higher experience levels. From the subjective feedback of participants, we also found comments approving of the unfamiliar slope plots and their inclination to adopt them as part of their own analysis workflow.

In another study that we conducted with a group of 36 climate scientists, we again found discrepancies between scientists' perceived and objective performance accuracy. While more than 60% of the scientists believed that they were more accurate with a familiar a rainbow color scale than with other alternatives, their average accuracy with the rainbow color scale in all three tasks in the study were less than the others. In fact, in the tasks involving magnitude estimation, they were significantly less accurate with the rainbow color scale as compared to the perceptually more optimal color scale. It also interesting to observe (Fig. 6.2) that compared to the much more skewed distribution with respect to familiarity, there is a much less skewed distribution with respect to preference - this implies that many participants were convinced after the study that the other optimal color scales were indeed better suited for the spatial data analysis tasks.

6.4 Critical Reflection

In this section, we summarize and reflect on the research questions that can be formulated based on the work discussed here. While some of the studies addressed these questions, they by no means provide a complete picture of the factors associated with familiarity related cognitive biases.

Does familiarity affect subjective impressions about visualization based judgments? We find evidence to believe that the familiarity heuristic has a strong effect on the preference for use of conventional visualizations by domain experts. However, participatory design sessions have proved to mitigate this effect [3].

A combination of participatory design and controlled experiments (Fig. 6.6) have helped understand the effects of the bias with a broader group of experts. Carefully conducted experiments, where experts have to conduct a set of tasks in controlled settings, have also been able to somewhat mitigate effects of the bias and indicate preferences in favor of the new, perceptually motivated visualizations. The reduction of bias was less prominent, however, for more experienced people who are potentially more hesitant in using and trusting the outcome of new analysis methods.

Does familiarity lead to better task performance? In our studies, we have used familiar visualizations which have been hypothesized to have certain shortcomings with respect to visualization design principles. In those cases, familiarity did not lead to a better performance. In fact, in most cases, experts performed better with

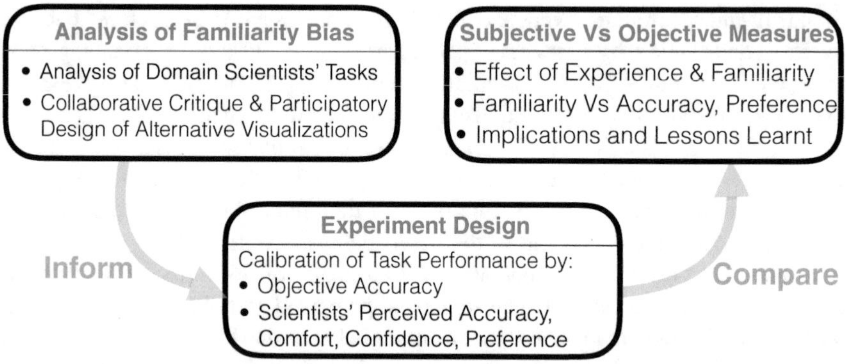

Fig. 6.6 A process framework for analyzing familiarity bias. The first step consists of a focused analysis of the bias and alternative visualization choices with a small group of experts. In some cases, participatory designs have proven an effective method for mitigating some of the biases and convincing experts to use better solutions. This is followed by a carefully designed experiment with ecologically valid tasks and metrics for recording experts' subjective ratings and objective performance. The final step is to analyze the study results for comparing subjective and objective performance measures

the unfamiliar visualizations, irrespective of their domain experience due to their perceptually optimal design. In the future, it will be interesting to compare familiar visualization techniques and systems with no shortcomings to unfamiliar ones to assess how strongly familiarity alone biases the judgment of experts.

How can biases associated with familiarity be measured? We used the discrepancy metric to understand how strongly the familiarity heuristic influences differences between perceived and objective performance measures. Metrics like persuasion [12] can also be used to evaluate if experts can be persuaded to not underestimate the detrimental effects of design problems. In cases where ground truth for expert judgments is unavailable, we can use a consensus metric to see if groups of experts agree or disagree about the decisions made based on visual evidence and understand the effects of the bias on population samples.

What are the implications for visualization adoption? A key decision for experts when performing their analysis is which visualization technique to use to address their tasks. As discussed earlier, the familiarity heuristic is a key determinant for this decision. While small case studies are able to demonstrate the effectiveness of new techniques, these are inadequate to change the lack of self-efficacy associated with adoption of new techniques. Interactive visualization techniques in many cases are disruptive for a domain. Sustained research collaborations are needed for longer-term adoption of such unfamiliar techniques and new metrics and studies need to be developed to judge the factors responsible for expert adoption [14].

6.5 Conclusion

In this chapter, we have provided a descriptive analysis of the strong role that the familiarity heuristic plays in experts' judgments using visualizations. The main problem with the familiarity heuristic is that it causes experts to subconsciously rely more on the conventional methods that might lead to sub-optimal performance. However, we have shown that through participatory design which carefully considers both experts' requirements and visualization design principles, we are not only able to inspire greater levels of experts' subjective preference in the alternative unfamiliar methods, but also find demonstrable evidence where perceptually motivated design can minimize the effect of familiarity, leading to greater performance accuracy.

These are still early research efforts in the direction of understanding biases related to familiarity. To make the use of visualization a viable and lasting solution for domain experts, we need to pursue the outlined research questions and work together towards efforts that lead to long-term adoption of *the best practices in* visualization methods and techniques.

References

1. Allen TT (2005) Introduction to engineering statistics and six sigma: statistical quality control and design of experiments and systems. Springer, Berlin
2. Ashcraft M (2006) Cognition
3. Brehmer M, Ng J, Tate K, Munzner T (2016) Matches, mismatches, and methods: multiple-view workflows for energy portfolio analysis. IEEE Trans Visual Comput Graphics 22(1):449–458
4. Dasgupta A, Poco J, Wei Y, Cook R, Bertini E, Silva CT (2015) Bridging theory with practice: an exploratory study of visualization use and design for climate model comparison. IEEE Trans Visual Comput Graphics 21(9):996–1014
5. Dasgupta A, Burrows S, Han K, Rasch PJ (2017a) Empirical analysis of the subjective impressions and objective measures of domain scientists' visual analytic judgments. In: Proceedings of the 2017 CHI conference on human factors in computing systems. ACM, New York, pp 1193–1204
6. Dasgupta A, Lee JY, Wilson R, Lafrance RA, Cramer N, Cook K, Payne S (2017b) Familiarity versus trust: a comparative study of domain scientists' trust in visual analytics and conventional analysis methods. IEEE Trans Visual Comput Graphics 23(1):271–280
7. Dasgupta A, Poco J, Rogowitz B, Bertini E, Silva CT (2018) The effect of color scales on climate scientists objective and subjective performance in spatial data analysis tasks. IEEE Trans Visual Comput Graphics (in publication)
8. Davis FD (1989) Perceived usefulness, perceived ease of use, and user acceptance of information technology. MIS Q 13:319–340
9. Gist ME, Mitchell TR (1992) Self-efficacy: a theoretical analysis of its determinants and malleability. Acad Manage Rev 17(2):183–211
10. Gleckler PJ, Taylor KE, Doutriaux C (2008) Performance metrics for climate models. J Geophys Res: Atmos 113(D6):D06104
11. Gulati R (1995) Does familiarity breed trust? the implications of repeated ties for contractual choice in alliances. Acad Manage J 38(1):85–112
12. Pandey AV, Manivannan A, Nov O, Satterthwaite M, Bertini E (2014) The persuasive power of data visualization. IEEE Trans Visual Comput Graphics 20(12):2211–2220

13. Park CW, Lessig VP (1981) Familiarity and its impact on consumer decision biases and heuristics. J Consum Res 8(2):223–230
14. Plaisant C (2004) The challenge of information visualization evaluation. In: Proceedings of the working conference on Advanced visual interfaces. ACM, New York, pp 109–116
15. Raskin J (1994) Intuitive equals familiar. Commun ACM 37(9):17–19
16. Samuelson W, Zeckhauser R (1988) Status quo bias in decision making. J Risk Uncertainty 1(1):7–59
17. Stajkovic AD, Luthans F (1998) Self-efficacy and work-related performance: a meta-analysis. Psychol Bull 124(2):240
18. Takayama L, Kandogan E (2006) Trust as an underlying factor of system administrator interface choice. In: Extended abstracts on Human factors in computing systems. ACM, New York, pp 1391–1396
19. Tversky A, Kahneman D (1973) Availability: a heuristic for judging frequency and probability. Cogn Psychol 5(2):207–232
20. Tversky A, Kahneman D (1991) Loss aversion in riskless choice: a reference-dependent model. Q J Econ 106(4):1039–1061

Chapter 7
Data Visualization Literacy and Visualization Biases: Cases for Merging Parallel Threads

Hamid Mansoor and Lane Harrison

7.1 Introduction

Data visualizations represent complex information to aid people in activities including exploration, analysis and decision-making. Given that visualizations rely on visual cues and involve factors like uncertainty and risk, data visualizations are prone to many perceptual and cognitive biases that humans are known to be susceptible to. These biases may lead people to come to the wrong conclusion from the data which may lead to poor decisions. Biases then in a sense may render data useless, as the relatively objective aspects of "data" is replaced with systematic and sometimes unpredictable results from biases. This problem is particularly an issue in today's world, given the growing role of data visualizations in people's day to day lives.

Data visualizations are used in everything from political analyses to consumer websites, financial tools and more. Given the sheer number and diversity of people who use visualizations, individual differences may have a large impact on how effectively viewers read data visualizations. One important individual difference in this respect is data visualization literacy, i.e. measures of how proficient people are at reading charts and graphs. There are unique challenges in measuring data visualization literacy, in part, due to the abundance of types of visualizations, the number of possible tasks that can be performed on a visualization and the actual data represented in the visualization. Even with fixed visualizations and tasks, there are challenges related to choosing an appropriate metric to represent literacy, such as a score, percentile, a grade, etc. Researchers have begun to address this gap, by developing and evaluating measures of data visualization literacy [4, 16].

H. Mansoor (✉) · L. Harrison
Worcester Polytechnic Institute, Worcester, MA, USA
e-mail: hmansoor@wpi.edu

L. Harrison
e-mail: ltharrison@wpi.edu

© Springer Nature Switzerland AG 2018
G. Ellis (ed.), *Cognitive Biases in Visualizations*,
https://doi.org/10.1007/978-3-319-95831-6_7

In parallel, there has been a growing interest and research developments at the intersection of biases and data visualization. Works in this area include initiatives such as identifying biases that manifest in visualizations [8–10], quantifying the impact of biases on visualization task performance and design expectations [19], and mitigating biases as they occur [11, 20]. In particular, there are calls for more attention on the methods and factors researchers use when evaluating biases in visualization, given these are still being uncovered and quantified (e.g. [8–10]).

The aim of this chapter is to make a case for merging the parallel threads of data visualization literacy and visualization biases. In doing so, we highlight research in cognitive biases [5, 18], focusing on studies which have established that cognitive ability and experience can play a role in how susceptible a person is to a particular type of bias. The results, research methods and organizational frameworks from these prior works may provide the visualization community with new means for investigating biases in data visualizations. For example by placing more emphasis on how variation in the impact of bias may be related to variations in human abilities such as their ability to inhibit biases, or by establishing that some biases are inevitable regardless of a person's individual experience and ability. Merging the data visualization literacy and visualization bias threads may also bring implications for visualization design, such as highlighting pitfalls for using more complex visualization types to mitigate biases, given that users with low visualization literacy or experience may have trouble using them.

To illustrate how visualization literacy and biases may interact, we revisit prior work on visualization biases. For example, we cover studies on the attraction bias and availability bias from Dimara et al. [9, 10], and discuss how data literacy measures could add dimensions and potentially impact their analyses and resulting discussions. We also cover studies that propose the use of visualizations to mitigate bias, such as Dragicevic et al. [11], and show how results in visualization literacy [16] may mediate their effectiveness. Taken together, these examples imply that as data visualization research continues to identify and quantify the biases that occur in visualizations, the impact of people's individual abilities may prove to be an important consideration for analysis and design.

7.2 Background

Beyond the information visualization community (e.g. [12, 13, 16]), prior work in visualization literacy spans communities such as intelligent tutoring systems [3] and K12 education [2, 14, 21]. The background discussed here covers work from these areas, in particular focusing on developments in visualization literacy that may relate to research targeting visualization biases.

7.2.1   Measuring Data Visualization Literacy and Quantifying Its Impact on Performance

Recent work in data visualization literacy has focused on the accurate assessment and representation of visualization proficiency. These measures are often accompanied by studies which illustrate the impact of high or low visualization literacy on tasks involving data visualizations.

Boy et al. [4] introduce a principled methodology for constructing assessment tests. Their methodology provides a blueprint for designers to devise comprehensive, scalable and rapidly deployable literacy assessment tests. They demonstrate the use of those rules in a user study containing four tests: two for line graphs and one each for bar charts and scatterplots. Having validated the predictive quality of this test, their work may be used by visualization researchers to add a literacy assessment component to their studies quickly and with little cost.

Lee et al. [16] propose the Visualization Literacy Assessment Test (VLAT), which leverages a six-step iterative process from Psychological and Educational Management research [6] alongside input from visualization experts to assign a numerical score of visualization literacy, with a specific focus on distinguishing expert visualization users from novices. An important consideration in their design was the range of possible tasks. Given a scatterplot, for example, participants are asked questions not only about individual points but also comparisons between points and trends. The VLAT's use of a range of visualization tasks helps it gauge a broad spectrum of participants' abilities with visualization. In a crowdsourced study, participants took the VLAT and then attempted a few questions about an unfamiliar data visualization, a Parallel Coordinates Plot (PCP). The results of this study indicated that participants who score high on the VLAT were more likely to perform well with a visualization unfamiliar to them.

Tests that assess visualization literacy may help designers gain an idea of how their target audience's proficiency aligns with their own, which can lead to more effective visualization designs. Additionally, researchers can readily add these tests to their on visualization experiments, with relatively little cost in terms of participant effort or analysis time. Future research surrounding the assessment of visualization literacy may continue to cover more visualization tasks and contexts, given that everyday people are viewing visualizations at a greater rate than ever before.

7.2.2   Novices, Experts and Visualization Use

Several studies have examined the thought process behind novices' interpretation and creation of data visualizations. Such studies aim to develop models to capture thought process to help visualization creators in their design process. While the studies reported here focus on novices and visualization use or construction, they also highlight potential biases that users face when having little experience with a visualization type or task.

Lee et al. [15] capture the novice thought process with the NOVIS model, which details five steps in which novices read data visualizations including encountering the visualization, constructing a frame, exploring the visualization, questioning the frame and floundering on the visualization. To develop NOVIS, they asked participants (college students) questions about three unfamiliar visualizations (parallel coordinates, chord diagrams and treemaps), followed by asking them to verbalize their approach as they navigated through the charts. Students' comments were then used to infer and generalize five stages of how novices approach unfamiliar charts.

Beyond visualization use and interaction, research has also focused on how novices create visualizations. A study from Grammel et al. [12] aimed to investigate the barriers that novices face when creating visualizations. In a user study, participants were asked to generate data visualizations through a mediator using Tableau and to verbalize their thoughts while generating the visualizations. Novices were reportedly unable to consistently specify visualizations and indicated that their preferences were influenced by their experience with prior data visualization types. The results of this think-aloud data gathering led to the proposal of three main barriers to visualization creation: selection, visual mapping and interpretation.

The education community has also studied how novices read charts. A study from Baker et al. [2] examined how K12 students interpreted and generated data visualizations. They presented middle school students with exercises to generate histograms, scatterplots and stem-and-leaf plots, capturing their design and construction process. Students reported little experience with these particular plots, but had considerably more experience with bar charts as a result of their schools' curriculum. The study found that the generation, interpretation and selection of the new visualizations were heavily influenced by the transfer of prior experience of bar charts. The researchers demonstrate that this bias may have been dependent upon surface similarities between bar charts and the other charts.

Kwon et al. [13] studied how effective different tutorial techniques are for teaching visualization skills. With the goal of having participants become proficient with parallel coordinates plots, they constructed multiple tutorials; a baseline condition contained no tutorial. A static tutorial included descriptions of parallel coordinates plots along with screenshots and a video tutorial showed participants how to navigate a PCP. Finally, an interactive tutorial allowed users to draw parallel coordinates, enter values and interact with the chart they made. It also gave users a list of tasks to complete, providing feedback when the users were unable to correctly finish a task. The results of a user study found that participants who saw the video and interactive tutorials fared better than the baseline and static tutorials. This study suggests that skill with a visualization can be learned and learned relatively quickly with proper training methods.

7.2.3 Biases and Data Visualization

People are prone to many types of biases when using data visualizations. Biases that manifest in data visualizations can impact a person's performance with visualization and possibly lead to errors in decision-making tasks related to the underlying data. This growing research area focusing on biases in visualization has investigated areas such as the mitigation of biases [8], assessing the impact and prevalence of specific biases [10] and developing approaches to analyze biases in visualizations [20]. Results from studies on bias in data visualizations can improve the ways in which visualizations are designed, benefiting the visualization community at large by enabling guidelines for less error-prone transfer of information.

7.3 Individual Differences and Bias: Guiding Results and Organizational Frameworks

Taken together, the threads of research in data visualization literacy and visualization biases have several parallels. Studies in data visualization literacy have uncovered biases that manifest through unfamiliarity with a visualization, for example, which may not happen when experts use the same visualizations. Beyond these threads, research in cognitive psychology has focused on the systematic study of the relationship between biases and individual differences. From the visualization perspective, Peck et al. discusses some possible implications of linking individual differences with factors such as experience and bias [17]. Here we highlight some of the extant research in biases and individual differences, focusing on results and organizational frameworks which may inform future studies in visualization literacy and bias.

In a series of experiments, Stanovich and West [18] studied the relationship between measures of cognitive ability and known biases. For cognitive ability, they adopted the SAT (Scholastic Aptitude Test) scores of their participants, who were primarily students. They used established bias experiments, including studies on base-rate neglect, anchoring effects, outcome bias. The results of their experiments indicated that some of these were uncorrelated with participants' cognitive ability. Others, however, did show an effect. To reconcile this difference and provide guidance for future experiments, Stanovich and West propose a "mindware" organizational framework which illustrates the ways in which ability may or may not impact performance in bias-prone tasks. This framework is shown in Fig. 7.1.[1] Their overall conclusion was that a person with high cognitive ability may be more able to take extra measures to prevent bias-induced errors if they are informed beforehand that the task they are about to perform involves a particular type of bias.

We know from extant research that viewing visualizations may result in many sorts of biases, but are all people equally susceptible? Stanovich and West [18] offer

---

[1]This image is a recreation of the framework in Stanovich et al. [18].

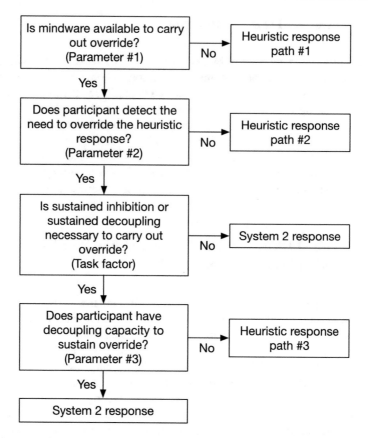

Fig. 7.1 Stanovich and West propose an organizational framework for reasoning about when individual differences may play a role in a person's ability to mitigate biases [18]. Given known effects of experience, ability, and bias, we propose that similar measures be adopted and used in the study of biases that occur in visualization use

examples of how long-term exposure and training with statistics and probabilities may equip people with mindware that allows altering-responses to be triggered, also known as a bias inhibiting response. As the visualization community continues to quantify the impact and ways of measuring visualization literacy, it is possible that long-term exposure to visualizations and deliberate practice with visualizations may result in people developing bias inhibiting responses for biases that occur when viewing data visualizations.

7.4 Linking Data Visualization Literacy to Existing Studies of Visualization and Biases

Given evidence from prior work that individual differences in visualization literacy can impact people's performance with data visualizations, that individual differences can play a role in the impact of biases and an organizational framework for thinking about this interplay [18], we now consider how results and methods from data visualization literacy research could be integrated into existing studies of visualization and biases.

THE AVAILABILITY BIAS: Dimara et al. [9] examined the availability bias and how it may manifest in visualization. They present a political voting decision as an example of a process that can fall prey to the availability bias. To mitigate the bias in this situation, they propose three ways in which data visualizations can help, focusing on how visualizations can aid recall to remove biases, and how heuristic inspired visualizations may be able to strike a balance between simplicity and accuracy to aid visualization users in avoiding biases. Following these mitigation strategies, they suggest that imperfections in visualizations can be tolerated if they increase understanding, a sentiment echoed by Correll and Gleicher [7]. What is less understood, however, is how factors like imprecise representations and complexity are related to variations in user ability. To give a concrete example, a complex visualization that mitigates the availability bias in the hands of an expert may be fine, but studies from Kwon et al. [13] note that novices enter many distinct stages when learning a new visualization. Thus, the populations used when testing visualization mitigation strategies should be taken into account, to ensure validated mitigation strategies perform as expected with potential end-users (Fig. 7.2).

THE ATTRACTION BIAS: In another study, Dimara et al. [10] examined the prevalence of the attraction bias in data visualizations using scatterplots and a table, as the control. The viewers were given a choice of two options, with an inferior

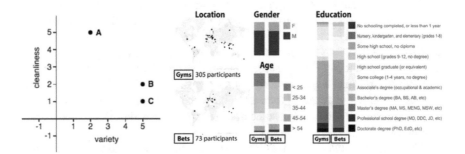

Fig. 7.2 (Left) Dimara et al. establish the attraction bias in visualizations such as scatterplots, where a decoy point can systematically bias participant choice. (Right) While their study includes measures of education and other individual differences, measures of cognitive ability have been shown to play a role in biases. Newly developed visualization ability assessments [4, 16] may add informative dimensions to bias studies such as these

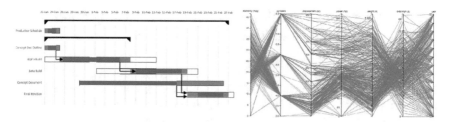

Fig. 7.3 (Left) Dragicevic et al. [11] propose visualizations like PlanningLines [1] (alternate implementation shown) as a means for mitigating the planning fallacy. (Right) However, Lee et al. found that the efficacy of more complex visualizations (such as parallel coordinate plots), are significantly modulated by visualization literacy scores [16], suggesting that the use of more complex visualizations may lead to additional performance costs

(and irrelevant) third choice, placed close to one of the others. The results indicated that data visualizations are also prone to this bias, causing people to make errors in judgment.

These study could benefit by adding literacy measures as a factor, especially as this study had participants of varying backgrounds and a literacy assessment test, such as those proposed by Boy et al. [4] or Lee et al. [16] could potentially uncover additional signals in the data. A numerical score of literacy could readily be factored into a correlation calculation and such an analysis may identify subpopulations that are more or less susceptible to attraction biases, or establish the robustness of the bias across varying backgrounds. A correlation between visualization literacy and biases may help designers create more effective visualizations for subpopulations that vary in literacy levels. Furthermore, frameworks for reasoning about individual differences and biases (e.g. Fig. 7.1) could be used in discussions to reason about whether mitigation strategies are possible for a given bias.

THE PLANNING FALLACY: In contrast to examining biases that manifest through visualization use, Dragicevic et al. [11] proposed using data visualizations to mitigate the planning fallacy, a common bias which occurs when people make time-lines for projects. They propose 4 ways in which visualizations may help prevent the planning fallacy. Namely, by providing, increased awareness of obstacles, self-logging of durations and predictions, tools for supporting group predictions, and social networking tools. Their discussion moves beyond individuals and into teams, as team projects may be more susceptible to planning fallacies, given that team members are often unaware of the each other's schedules and skills. As a possible mitigating visualization meeting some of these criteria, they describe Aigner et al.'s PlanningLines [1] (shown in Fig. 7.3).[2]

The interplay between complex visualizations and visualization literacy could apply in this case. Specifically, Lee et al. [16] showed in their study evaluating the Visualization Literacy Assessment Test (VLAT) that high scores were positively associated with peoples' ability to navigate unfamiliar visualizations. While the VLAT

---

[2]PCP source: https://bl.ocks.org/jasondavies/1341281.

study focused on PCPs, the proposed PlanningLines visualization uses a variety of visual encodings - glyphs, links, overlapping bars for uncertainty, etc., which could imply that experience or training are necessary to achieve the goal of mitigating the planning fallacy. PlanningLines is an unfamiliar visualization and may cause issues for novices in a group who misinterpret it. Further, as the tool is proposed to be used in teams, it is unknown how individual differences such as visualization literacy will manifest when people with different abilities are modifying and interpreting the same visualizations.

7.4.1 *Reversal: Augmenting Data Visualization Literacy Research with Biases*

So far, we have discussed how incorporating literacy as a factor in studies concerned with biases in visualizations may lead to deeper or different conclusions. However, we also note that the ongoing development of visualization literacy assessment approaches can be informed by research in visualizations and biases. Broadly speaking, visualization literacy is a measure of the ability to accurately perform basic chart reading tasks, which may be compromised due to biases. Specifically, if it can be shown that certain biases that manifest in visualizations can be inhibited through experience and training, assessment questions that include such bias-prone tasks may prove useful in discriminating between novice and experienced users.

7.5 Conclusion

The aim of this position chapter is to make a case that research in data visualization literacy and biases in visualization are two parallel threads that, when merged, may reveal new insights about one another. In making this case, we draw on work from cognitive psychology that has established that individual differences and biases are often (though not always) related, and discuss one of the resulting organizational frameworks in relation to thinking about the conditions under which visualization biases might be mitigated. We revisit several studies on visualizations and biases and tie them to extant works in data visualization literacy. As these related areas continue to grow, mutual consideration may prove beneficial in furthering our understanding of visualization analysis and design.

References

1. Aigner W, Miksch S, Thurnher B, Biffl S (2005) Planninglines: novel glyphs for representing temporal uncertainties and their evaluation. In: Proceedings. Ninth international conference on Information visualisation, 2005. IEEE, pp 457–463

2. Baker RS, Corbett AT, Koedinger KR (2001) Toward a model of learning data representations. In: Proceedings of the 23rd annual conference of the Cognitive Science Society, pp 45–50

3. Baker RS, Corbett AT, Koedinger KR (2004) Learning to distinguish between representations of data: a cognitive tutor that uses contrasting cases. In: Proceedings of the 6th international conference on Learning sciences, International Society of the Learning Sciences, pp 58–65

4. Boy J, Rensink RA, Bertini E, Fekete JD (2014) A principled way of assessing visualization literacy. IEEE Trans Visualization Comput Graphics 20(12):1963–1972

5. Christensen-Szalanski JJ, Beck DE, Christensen-Szalanski CM, Koepsell TD (1983) Effects of expertise and experience on risk judgments. J Appl Psych 68(2):278

6. Cohen RJ, Swerdlik ME, Phillips SM (1996) Psychological testing and assessment: An introduction to tests and measurement. Mayfield Publishing Co

7. Correll M, Gleicher M (2014a) Bad for data, good for the brain : Knowledge-first axioms for visualization design. In: Ellis G (ed) DECISIVe : Workshop on Dealing with Cognitive Biases in Visualisations, http://nbn-resolving.de/urn:nbn:de:bsz:352-0-329455

8. Correll M, Gleicher M (2014b) Error bars considered harmful: exploring alternate encodings for mean and error. IEEE Trans Visualization Comput Graphics 20(12):2142–2151

9. Dimara E, Dragicevic P, Bezerianos A (2014) Accounting for availability biases in information visualization. In: Ellis G (ed) DECISIVe : workshop on dealing with cognitive biases in visualisations, http://nbn-resolving.de/urn:nbn:de:bsz:352-0-329436

10. Dimara E, Bezerianos A, Dragicevic P (2017) The attraction effect in information visualization. IEEE Trans Visualization Comput Graphics 23(1):471–480

11. Dragicevic P, Jansen Y (2014) Visualization-mediated alleviation of the planning fallacy. In: Ellis G (ed) DECISIVe : workshop on dealing with cognitive Biases in visualisations, http://nbn-resolving.de/urn:nbn:de:bsz:352-0-329469

12. Grammel L, Tory M, Storey MA (2010) How information visualization novices construct visualizations. IEEE Trans Visualization Comput Graphics 16(6):943–952

13. Kwon BC, Lee B (2016) A comparative evaluation on online learning approaches using parallel coordinate visualization. In: Proceedings of the 2016 CHI conference on human factors in computing systems, ACM, pp 993–997

14. Laina V, Wilkerson M (2016) Distributions, trends, and contradictions: a case study in sensemaking with interactive data visualizations. In: Proceedings of the 12th international conference of the learning sciences. The international society of the learning sciences, Singapore

15. Lee S, Kim SH, Hung YH, Lam H, Ya Kang, Yi JS (2016) How do people make sense of unfamiliar visualizations?: a grounded model of novice's information visualization sensemaking. IEEE Trans Visualization Comput Graphics 22(1):499–508

16. Lee S, Kim SH, Kwon BC (2017) Vlat: development of a visualization literacy assessment test. IEEE Trans Visualization Comput Graphics 23(1):551–560

17. Peck EM, Yuksel BF, Harrison L, Ottley A, Chang R (2012) Position paper: towards a 3-dimensional model of individual cognitive differences. ACM BELIV

18. Stanovich KE, West RF (2008) On the relative independence of thinking biases and cognitive ability. J Personality Social Psych 94(4):672

19. Tenbrink T (2014) Cognitive discourse analysis for cognitively supportive visualisations. In: Ellis G (ed) DECISIVe : workshop on dealing with cognitive biases in visualisations, http://nbn-resolving.de/urn:nbn:de:bsz:352-0-329485

20. Verbeiren T, Sakai R, Aerts J (2014) A pragmatic approach to biases in visual data analysis. In: Ellis G (ed) DECISIVe : workshop on dealing with cognitive biases in visualisations, http://nbn-resolving.de/urn:nbn:de:bsz:352-0-329490

21. Wilkerson-Jerde MH, Wilensky UJ (2015) Patterns, probabilities, and people: making sense of quantitative change in complex systems. J. Learning Sci 24(2):204–251

Chapter 8
The Biases of Thinking Fast and Thinking Slow

Dirk Streeb, Min Chen and Daniel A. Keim

8.1 Introduction

Daniel Kahneman's "Thinking, Fast and Slow" [8] is a best-selling book about the biases of human intuition. The book provides an insightful and enjoyable omnibus tour of human mind in decision-making, drawing from empirical findings in a wide range of research publications in cognitive sciences and especially from the research activities of the author, Amos Tversky, and their colleagues and students. As one of the most popular non-fiction books in this decade, it has generated a profound impact on politics, business, healthcare, and many scientific and scholarly fields.

In "Thinking, Fast and Slow", Kahneman describes two *systems* of decision-making. "*System 1* operates automatically and quickly, with little or no effort and no sense of voluntary control. *System 2* allocates attention to the effortful mental activities that demand it, including complex computations." [8, p. 20]. A major contribution of Kahneman's scientific work was to point out numerous differences between prescriptions of normative theory and observed human behaviors. The book highlights more than twenty types of such differences as weaknesses of System 1 thinking, e.g. priming effect (§4), overestimating and overweighting rare events (§30), inference without statistics (§10, §16) or algorithms (§21). As many decisions made by System 2 may feature some shortcomings as well, Kahneman attributes these to System 2 being busy, depleted or lazy (§3).

D. Streeb (✉) · D. A. Keim
Data Analysis and Visualization Group at the University of Konstanz,
Konstanz, Germany
e-mail: streeb@dbvis.inf.uni-konstanz.de

D. A. Keim
e-mail: keim@dbvis.inf.uni-konstanz.de

M. Chen
Oxford e-Research Centre at the University of Oxford, Oxford, UK
e-mail: min.chen@oerc.ox.ac.uk

© Springer Nature Switzerland AG 2018

G. Ellis (ed.), *Cognitive Biases in Visualizations*,
https://doi.org/10.1007/978-3-319-95831-6_8

In the book, Kahneman champions statistics and algorithms (e.g. §16, §21). This view is often used as a supporting evidence for minimizing humans' role in data intelligence. From the perspective of visual analytics, which advocates the integration of machine-centric processes (i.e. statistics, algorithms and machine learning) with human-centric processes (i.e. visualization and interaction), the book appears to lean heavily towards machine-centric processes.

The labeling of Kahneman's two systems is meant to be a characterization rather than a panacea [20]. Similar dichotomies can be found in other psychological literature (cf. [10]). There are also suggestions for one-system or k-system models (e.g. [10]). In this paper, we follow the two-system discourse by considering two types of human-centric processes: (A) intuition-based decision-making and (B) analysis-based decision-making using, e.g. logical, statistical, rule-based and procedural reasoning. Later in Sect. 8.4, we add two additional types involving machine-centric processes: (C) fully automated decision-making using only machine-centric processes, and (D) *visual analytics* where machine- and human-centric processes are integrated together for decision-making.

In the remainder of this chapter, we first summarize the scholarly discourse on heuristics and biases in the literature (Sect. 8.2). We then examine three case studies in Kahneman's "Thinking, Fast and Slow" [8], pointing out the potential biases that Kahneman may have unwittingly introduced (Sect. 8.3). Finally we provide an alternative interpretation to the empirical findings in the book, making a case that the visual analytics paradigm remains to be a necessary approach to many complex decision processes (Sect. 8.4).

8.2 Heuristics and Biases

In statistics, *bias* has a relatively clear definition, i.e. the expected value of a procedure A used for fitting or predicting a random variable B is not equal to the expected value of B. On the other hand, the notion of bias as used in the context of human decision-making is much looser. Here, bias describes the deviation of human decision-making from some option an experiment designer considers as being optimal. Thereby, bias is attributed a clearly negative meaning, which is not as prominent in statistics, where usually other measures are minimized, such as mean squared error.

In psychology and decision sciences, *heuristics* are commonly considered as decision strategies that enable fast and frugal judgments by ignoring some of the available information, typically under the conditions of limited time and information [6]. This notion characterizes the Type (A) human-centric processes. One earlier and different notion of *heuristic* is that they are useful strategies to cope with problems that are intractable using exact computation such as playing chess [18]. This implies the Type (B) human-centric processes, and corresponds to some machine-centric processes such as the concept of *heuristic algorithms* in computer science. Regardless of which notion, psychologists all agree that heuristics may lead to *biases*, resulting in various types of errors in judgments and decisions [8]. Hence heuristics and biases are the two sides of the same coin.

The research and discourse on heuristics and biases has attracted much attention in the literature. Fiedler and von Sydow [5] provide an overview of the research in psychology since Tversky and Kahneman's 1974 paper [21]. There are papers that focus on specific examples in Kahneman's book [8], such as on the hot-hand fallacy [7, 13], the priming effect [24] and experiment reproducibility [9, 14, 17, 22].

Here we examine the sources of biases from a probabilistic perspective. As illustrated in Fig. 8.1, let us consider a domain of entities (e.g. objects, phenomena, or events) that is referred to as the *global domain*. Any systematic sampling of this domain is usually limited to a pre-defined set of variables. To maintain the statistical validity of the sampling process, the required sample size is exponentially related to the number of variables, which is understandably kept low in most situations. Meanwhile, an individual most likely encounters some of these entities in a sub-domain (referred to as a *local domain*). The observations made in these encounters typically include more variables (as illustrated by 3-D features and shape variations in Fig. 8.1) than what a systematic sampling may cover. To process a sequence of input data, Type (A) human-centric processes rely on the experience built on accumulated information about the local domain, while Type (B) human-centric processes rely on the summary resulting from systematic sampling in the global domain. In addition, both will make use of some a priori knowledge, though the two will likely utilize or focus on different forms of knowledge.

Many shortcomings of human heuristics highlighted by Kahneman [8] represent attempts of using experience about a local domain to make a decision about an input from the global domain. In such situations, it is often reasonable to think that using the global statistics would be more appropriate. However, if a decision is to be made about an input that is known to be in a local domain and features more variables than those covered by the global sampling, using the experience accumulated in this local domain could be more appropriate than blindly using the global statistics.

Both Kahneman [8, p. 222] and Gigerenzer [6, p. 4] point out that more information is not necessarily the key to better decisions. Instead, it is essential to choose the more appropriate global and/or local information in a given situation. In the next section we use some of Kahneman's examples to illustrate that biases can result from inappropriate use of global or local information.

8.3   Case studies

We appreciate that most of the case studies in Kahneman's book [8] are used to illustrate the biases of human intuition, and the absolute necessity for anyone working in data science and decision science to be aware of the limits posed on decision-making processes. We find that most of his case studies are well designed and presented. Here we only show three case studies that feature biases unaccounted for in the book. We flag these cases to illustrate that the sources of biases may come from both sampling domains as shown in Fig. 8.1. One should not read the discourse on these case studies as an argument against the main thesis of Kahneman's book.

8.3.1 Causes Trump Statistics

In §16 Causes Trump Statistics, Kahneman describes the following synthetic scenario:

> "A cab was involved in a hit-and-run accident at night. Two cab companies, the Green and the Blue, operate in the city. You are given the following data: • 85% of the cabs in the city are Green and 15% are Blue. • A witness identified the cab as Blue. The court tested the liability of the witness under the circumstances that existed on the night of the accident and concluded that the witness correctly identified each one of the two colors 80% of the time and failed 20% of the time." [8, p. 166, ch. 16]

Using Bayesian inference, Kahneman concludes that the probability of the guilty cab being Blue is 41%, though most people would instinctively ignore the base-rate of 15% and go with the witness for 80%. This example is meant to illustrate a failure of human reasoning.

It is necessary to notice the difference between the sampling domains involved. The overall statistics of cabs in the city are likely to be compiled in the domain of the whole city for all day and night. However, the incident happened in a particular area in the city and at a night. There is no reason to ascertain that the global statistics would be the same as the local statistics. This is of the same logic that the local sample cannot simply be assumed to be representative of the global population. Meanwhile, the court is likely to test the reliability of the witness indoors using two-dimensional photos. This sampling domain is different from the outdoor environment where the witness observed a real three-dimensional car. In addition, when one watches a real world event unfolding, one observes many variables, such as the color and shape of the cab and the telephone number and the font in which it is written on the cab. As memorization and recall involve an abstraction process, naturally one may use "Blue" as a cue for the memory representation. Hence, the witness is likely to remember more than just a color. Those additional variables are not part of the uni-variate global statistics about cab colors.

While it is reasonable to suggest that the global statistics about cabs should be considered at court, it is necessary to examine if it is possible for the local statistics to deviate from the global statistics. It is inappropriate to apply the indoor testing results (as in one local domain) to an outdoor scenario (a different local domain) without any moderation. Therefore, it is unsafe to simply combine the two pieces of statistics using Bayes' rule, and to conclude that the witness is likely to be wrong. Indeed, those people who trusted the witness may have taken the factors of sampling domain and additional variables into account. As a result this example is not necessary a case of "biases trump powerful statistics" but a case of "humans' heuristics trump biased statistics".

The potential deviation of statistical measures of a local sample from those of a global sample created by amalgamating many local samples is understood by statisticians. The well-known Simpson's paradox is such an example in a manifest form [16, 19, 23]. There are many similar real-world examples, one of which is a report by Bickel et al. on sex bias in university admission [2]. Furthermore, Birnbaum [3] points out that a normative analysis following Kahneman is incomplete due to its ignorance to signal detection theory and judgement theory. Krynski and Tenenbaum [11] argue that the problem features a high false-positive rate and a lack of a causal structure. In a later work [12], they highlight the overlooked value of humans' "causal knowledge in real-world judgment under uncertainty" [15]. While there are also papers (e.g. [1]) that support Kahneman's conclusion that Bayesian inference is superior to humans' intuition, they are all based on the crucial but subjective assumptions that the problem did not feature a partial Simpson's paradox and the witness did not observe extra information in addition to colors.

Taking all these points together, we can observe that while the case study was presented as an illustration of humans' bias in decision-making, its Bayesian inference also features biased assumptions. Were these biased assumptions properly taken into consideration, the conclusion could be very different.

Relevance to Visualization. Interactive visualization typically provides indispensable support to spatiotemporal data analysis. It is usually difficult to capture all necessary variables that describe the characteristics of each location and each path between two locations. Any statistical aggregation across many locations and routes can thus be very sensible to uncaptured confounding variables. Visualization stimulates human analysts' knowledge about these uncaptured variables, while interaction allows analysts to explore different spatial regions and temporal ranges of the data according to both global statistics as well as local characteristics.

8.3.2   Tom W's Specialty

In the book, Kahneman presents another case study in favor of a base-rate (i.e. a piece of global statistics).

> "Tom W is a graduate student at the main university in your state. Please rate the following nine fields of graduate specialization in order of the likelihood that Tom W is now a student in each of these fields. [...] • business administration • computer science [...] • social science and social work" [8, p. 146, ch. 14]

Kahneman further describes a personality sketch of Tom W. He then states that the variables featured in the personality sketch (e.g. nerdiness, attention to details, sci-fi type, etc.) are irrelevant to the question, and only the base-rate of enrollments in these subjects provides the solution.

Most of us agree that for Tom W to choose a subject to study is a complicated decision process. This may depend on many variables, including some captured in his personality sketch, as well as many others unknown to readers. As illustrated

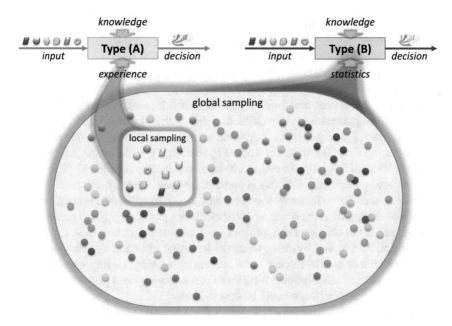

Fig. 8.1 The two types of human-centric processes typically correspond to the information acquired from, and knowledge related to, two sampling domains

in Fig. 8.1, the base-rate is a piece of global statistics based on the variable of enrollment outcomes, while the personality sketch define a local sampling domain for those fitting the sketch. Ideally one would also wish to have other global statistics, such as enrollment outcomes according to some twenty variables mentioned in his personality sketch (e.g. gender, intelligence, creativity, nerdiness, attention to details, writing skills and so on), since many of these are confounding variables that influence the base-rate.

Understandably, most organizations do not have all these global statistics or feel inappropriate to disclose them. Nevertheless, using only the base-rate of one variable to infer a clearly multi-variate decision will introduce a huge amount of bias against the conditional probability constrained by the local sampling domain. Such bias is alarmingly dangerous, for instance, by imagining that personality sketches were ignored in criminal investigation, business management, and the media and entertainment industries. In practice, the lack of comprehensive global statistics and the biases of uni-variate approximation of a multi-variate problem is usually compensated by human heuristics that have captured some multi-variate information of a local sampling domain.

Notably, Kahneman himself points out that the probability of Tom W's subject is complex. "Logicians and statisticians disagree about its meaning, and some would say it has no meaning at all." [8, p. 150, ch. 14].

Let "Tom studies $a \in X$" be a truth statement, where X is a variable of subjects and a is one particular subject. The probability function $P(X \mid \text{Tom})$ is an approximation of this truth statement. Since a personality sketch is used to describe a set of variables of a person, $V_1, V_2, \ldots, V_n$, $P(X \mid V_1, V_2, \ldots, V_n)$ is a further approximation. While ignoring the base-rate $P(X)$ is clearly a naive bias, ignoring the impact of variables $V_1, V_2, \ldots, V_n$ is also a dangerous bias. For example, it is well known that, nowadays, women are overrepresented in psychology and underrepresented in computer science. Only by considering that Tom is male, can one correct the bias of the global sample towards psychology and away from computer science. Thus, in contrast to Kahneman, we consider $P(X \mid \text{Tom})$ to be the best approximation of the truth statement, not $P(X)$.

Relevance to Visualization. Multivariate data visualization techniques, such as parallel coordinates plots and glyph-based visual encoding, enable human users to observe many variables simultaneously. These techniques complement major statistical measures (e.g. base-rate) by preventing any complex decision process (e.g. ranking countries' healthcare quality) from simply depending on the base-rate of one variable (e.g. mortality rate).

8.3.3   The 3-D Heuristic

Let us consider a visual case study that Kahneman discussed in the early part of the book. After presenting a Ponzo illusion similar to Fig. 8.2a, Kahneman asks:

> "As printed on the page, is the figure on the right larger than the figure on the left?" [8, p. 100, ch. 9]

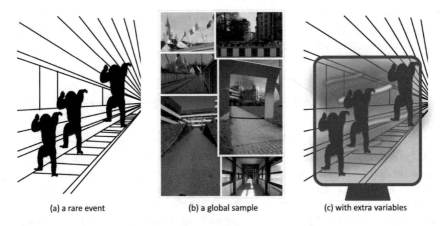

(a) a rare event   (b) a global sample   (c) with extra variables

Fig. 8.2 a An illusion redrawn based on one presented by Kahneman [8, p. 100, ch. 9], **b** examples of the global domain, **c** The illusion example **a** with additional information, associating the three figures with a 2D plane

Kahneman considers that the question is not misleading. "The bias associated with the heuristic is that objects that appear to be more distant also appear to be larger on the page." [8, p. 101]

We do agree that the Ponzo illusion demonstrates a bias in human perception. However, such bias does not in any way suggest that the humans' 3-D heuristic is ineffective. The humans' 3-D heuristic, in fact, is a form of reasoning using a base-rate, which is obtained by observing a huge number of instances in global sampling as illustrated in Fig. 8.2b. Illusions are outlier events, most of which were created purposely to illustrate the existence of instances contradicting the perception based on the base-rate.

In many cases, biases due to global statistics occur in machine-centric processes. In this case study, the base-rate biases occur in a human-centric process. Instead of considering this is a misdirected application of global statistics to a very specialized local domain, Kahneman somehow attributes the problem to human heuristics in general.

When applying global statistics to a local sampling domain, biases can often be alleviated by introducing additional variables that humans can reason with. In Kahneman's question, the phrase "As printed on the page" is designed to be misleading in order to maximize the illusion. Hence, the variable about whether the three figures are meant to be measured in their 2-D projections or the perceived 3-D reality is stated ambiguously. If the requirement of 2-D projections was clearly defined, e.g. as 2-D stickers on a piece of glass as illustrated in Fig. 8.2c, the illusion would become much less effective.

Relevance to Visualization. Gestalt grouping, which is a major cause of illusion, holds an important key to the power of visualization. If the human vision system was not equipped with Gestalt grouping, our ability to spot patterns, clusters, anomalies and changes in visualizations would be significantly degraded. But, it is a prerequisite for Gestalt grouping that humans read beyond the data provided by their eyes. Visualization designers should not worry about the existence of false readings, instead, they should reckon how large negative effects might be and how difficult it is to become conscious about them.

8.4   Implications for Visualization and Visual Analytics

The discourse in Sects. 8.2 and 8.3 shows that both types of human-centric processes, namely (A) human heuristic and (B) structured reasoning using logic, statistics, rules, and procedures, can lead to biases. If a local domain has the same variables and statistics as the global domain, i.e. being representative, both types of processes will likely function well in both domains. When they have different variables and statistics, for humans to perform decision-making tasks in the global domain will lead to biases if there is not enough context knowledge to build a bridge. Similarly, decision-making based on global statistics in a local domain can lead to biases if it does not encapsulate the variations and variables necessary for sound decisions in the local domain [4]. Hence, both approaches may suffer from biases.

A common desire is to replace both human-centric processes by (C) a fully automated decision system, which is equipped with many types of global statistics, logic, rules and algorithms. Kahneman's book [8] naturally inspires such an approach. The main challenges for such machine-centric processes to be effective are (i) the difficulty to collect enough data, in every local domain, to know if its sample features the same statistical measures as the global sample, (ii) the difficulty to sample a variety of variables as humans would observe in a real-world environment, and (iii) the difficulty in determining what variables should be used in a decision process, dynamically, according to contextual conditions. With rapidly increasing capabilities of data collection, some machine-centric processes will become more effective decision tools in individual local domains at an acceptable level of biases. However, we must beware of these biases due to global statistics and missing variables in global sampling.

For almost all complex decisions, the most appropriate approach is (D), namely, to support human-centric processes with machine-centric processes and vice versa. Following this approach, the two types of human-centric processes (A) and (B) are combined with machine-centric processes based on statistics, logic, rules and algorithms. But unlike (C), (D) does not intend to fully replace (A) and (B). Instead, (D) uses computers' extensive memory capacity and fast computation capability to assist humans and relieve (A) and (B) from computationally intense tasks. One most effective form of such help is *visualization*, which enables humans to observe a vast amount of raw data, statistics, analytical results, and computed predictions and recommendations, in a rapid fashion.

Meanwhile, the human in the loop of (D) is aware of its limitation in gathering all necessary variables and data for a decision, and its inability to judge whether it is biased towards local statistics. (D) allows human-centric processes to override automated decision recommendations. This is indeed the role of *Interaction* for humans to add their information and knowledge to decision processes.

(D) is what the field of visual analytics has been advocating. With (D) or *Visual Analytics*, semi-automated analyses (including statistics, logic, rules and algorithms) provide human analysts with (i) numerically aggregated information based on large volumes of data, that humans themselves cannot store, (ii) consistent and rapid application of logic, rules, and algorithms, which humans cannot match in speed and consistency, (iii) visual representations of a variety of information, that humans cannot easily imagine in their minds, and (iv) systematically sequenced interactions, which help structure the operational workflow. At the same time, this allows human analysts to inform (D) about information related to the tasks at hand and local variables known only to analysts as well as to guide (D) to process and display information that is most relevant to tasks and local contexts, alleviating the biases of global statistics and the logic, rules, and algorithms designed for common scenarios.

In summary, the studies of heuristics and biases in psychology have provided a rich collection of empirical evidence that enables us to appreciate the benefit of heuristics and the common existence of biases. We examine the sources of biases from a probabilistic perspective, revealing that the fundamental cause is the difference between the two sampling domains, i.e. the domain that human heuristics are based on or statistical samples are obtained from, and the domain heuristics and statistics are applied to for decision-making. Biases can be caused by both heuristics and statistics, or in general, by the information and knowledge related to both sampling domains as shown in Fig. 8.1. All this points to the conclusion that visual analytics, or (D), is the most sensible approach to complex decision-making.

Acknowledgements Part of this work was funded by the German federal state's Initiative of Excellence via the Graduate School of Decision Sciences at the University of Konstanz.

References

1. Bar-Hillel M (1980) The base-rate fallacy in probability judgements. Ars Psychologica 44(3):211–233
2. Bickel PJ, Hammel EA, O'Connell JW (1975) Sex bias in graduate admissions: data from Berkeley. Science 187(4175):398–404
3. Birnbaum MH (1983) Base rates in Bayesian inference: signal detection analysis of the cab problem. Am J Psych 96(1):85–94
4. Fiedler K (2008) The ultimate sampling dilemma in experience-based decision making. J Exp Psych: Learning, Memory, Cognition 34(1):186–203
5. Fiedler K, von Sydow M (2015) Heuristics and biases: beyond Tversky and Kahneman's (1974) judgment under uncertainty. In: Eysenck MW, Groome D (eds) Cognitive psychology: revisiting the classic studies, Sage Publications, chap 12, pp 146–161, https://www.worldcat.org/isbn/9781446294475
6. Gigerenzer G, Hertwig R, Pachur T (eds) (2011) Heuristics: the foundations of adaptive behaviour. Oxford University Press, New York, https://www.worldcat.org/isbn/9780190494629
7. Green B, Zwiebel J (2017) The hot-hand fallacy: cognitive mistakes or equilibrium adjustments? evidence from major league baseball. Manage Sci https://doi.org/10.2139/ssrn.2358747
8. Kahneman D (2011) Thinking, fast and slow. Penguin Books Ltd., London, https://www.worldcat.org/isbn/9781846140556
9. Kahneman D (2017) Comment in response to Schimmack et al. online, https://replicationindex.wordpress.com/2017/02/02/reconstruction-of-a-train-wreck-how-priming-research-went-of-the-rails/comment-page-1/#comments. Accessed in June 2017
10. Keren G, Schul Y (2009) Two is not always better than one. Perspect Psychological Sci 4(6):533–550
11. Krynski TR, Tenenbaum JB (2003) The role of causal models in reasoning under uncertainty. In: Proceedings of the 25th annual conference of the cognitive science society. Erlbaum, Mahwah, NJ. https://www.worldcat.org/isbn/9780805849912
12. Krynski TR, Tenenbaum JB (2007) The role of causality in judgment under uncertainty. J Exp Psych: Gen 136(3):430–450
13. Miller JB, Sanjurjo A (2016) Surprised by the gambler's and hot hand fallacies? A truth in the law of small numbers. IGIER Working Paper No 552 https://doi.org/10.2139/ssrn.2627354
14. Open Science Collaboration (2015) Estimating the reproducibility of psychological science. Science 349(6251): https://doi.org/10.1126/science.aac4716
15. Pearl J (2009) Causality: Models, reasoning and inference, 2nd edn. Cambridge University Press, https://doi.org/10.1017/CBO9780511803161
16. Pearl J (2013) Understanding simpson's paradox. Tech. rep., http://ftp.cs.ucla.edu/pub/stat_ser/r414.pdf

17. Schimmack U, Heene M, Kesavan K (2017) Reconstruction of a train wreck: how prim-
    ing research went off the rails. online, https://replicationindex.wordpress.com/2017/02/02/
    reconstruction-of-a-train-wreck-how-priming-research-went-of-the-rails/. Accessed in June
    2017
18. Simon HA (1990) Invariants of human behavior. Annu Rev Psych 41:1–20
19. Simpson EH (1951) The interpretation of interaction in contingency tables. J R Stat Soc Ser B
    (Methodological) 13(2):238–241
20. Sloman SA (1996) The empirical case for two systems of reasoning. Psychological Bulletin
    119(1):3–22
21. Tversky A, Kahneman D (1974) Judgment under uncertainty: heuristics and biases. Science
    185(4157):1124–1131
22. Wagenmakers EJ, Zwaan RA (2016) Registered replication report: Strack, Martin, & Stepper
    (1988). Perspect Psychological Sci 11(6):917–928
23. Wagner CH (1982) Simpson's paradox in real life. Am Statistician 36(1):46–48
24. Yong E (2012) Nobel laureate challenges psychologists to clean up their act: Social-priming
    research needs "daisy chain" of replication. https://doi.org/10.1038/nature.2012.11535, includ-
    ing a public letter by Daniel Kahneman

Part III
Mitigation Strategies

Chapter 9
Experimentally Evaluating Bias-Reducing Visual Analytics Techniques in Intelligence Analysis

Donald R. Kretz

> *Whether we say that the biases are 'irrational' is of no consequence. If we can help people make better judgments, that is a good thing.* Jonathan Baron [1].

9.1 Introduction

Errors in judgment in the national security domain can have disastrous consequences for entire regions of the world, perhaps even leading to wars between nations based on erroneous information. In a series of U.S. Government studies and reports since 2001, *cognitive biases* were identified as a major cause of analytic pathologies in the U.S. Intelligence Community [2–7]. Research shows that even trained, experienced and deeply committed personnel sometimes err in judgment as a result of cognitive heuristics. Despite this knowledge, there is still no "cookbook" of rigorously tested techniques to avoid judgment biases when solving complex, uncertain and ambiguous intelligence problems. For this reason, cognitive bias mitigation has become a focus of attention. There is still a need for experimentally validated debiasing techniques that can be incorporated into analytic tradecraft so that foreseeable thinking errors can be avoided. This chapter presents a brief discussion of the intelligence analysis process and a short history of cognitive bias mitigation in this highly specialized profession.

D. R. Kretz (✉)
Security Engineering and Applied Sciences, Applied Research Associates, Inc, Frisco, TX, USA
e-mail: dkretz@ara.com

D. R. Kretz
School of Behavioral and Brain Sciences, University of Texas at Dallas, Richardson, TX, USA

© Springer Nature Switzerland AG 2018
G. Ellis (ed.), *Cognitive Biases in Visualizations*,
https://doi.org/10.1007/978-3-319-95831-6_9

Developing effective debiasing techniques requires addressing a number of daunting challenges. While intuitively appealing, the ability to construct suitable methods to test behavior under actual work conditions is often limited. As often happens in research, the generalizability of findings from laboratory settings to work environments is a serious limitation. Although final judgments and conclusions are usually highly visible, decision processes are latent and cognitive errors can only be inferred from work artifacts or through self-described thought processes. Despite these challenges, a handful of researchers have performed investigations of a small number of debiasing techniques under rigorous testing conditions. Those studies are described later in this chapter, followed by a reference framework for studying bias mitigation techniques under experimental conditions.

9.2 Basics of Intelligence Analysis

Intelligence analysis is a complex process that requires substantial training and deeply honed expertise.[1] Intelligence analysts must solve complex problems involving incomplete and ambiguous data and must often do so under tremendous time pressure. Beyond the simple reporting of facts, analysts must draw conclusions, construct explanations and make predictions. To perform effectively, an analyst must apply techniques learned from professional training; simultaneously perform deductive, inductive and abductive reasoning; and integrate previously-acquired subject matter knowledge. At the same time, they must constantly assess the veracity of sources, the relevance of new information and the accuracy of prior analyses. A number of external factors combine to complicate the analytic process. Analyses are often highly perishable; a decision maker's window of opportunity may be extremely narrow. Adding further pressure to the process, the amount of time allocated to a particular analytic task is sometimes determined by a policy requirement or managerial deadline rather than by the nature or complexity of the problem being examined. Senior policy makers, military commanders and the public at large depend on the judgments made by intelligence organizations—and count on them being based on the best analysis of all available information, objective and bias-free.

9.2.1 Intelligence as a Cognitive Process

Studies performed during the 1970s at the U.S. Army Research Institute for the Behavioral and Social Sciences determined that intelligence analysis was an amalgamation of cognitive functions and concluded that any attempts to understand or

---

[1] For detailed material describing the intelligence analysis profession and its analytic tradecraft and methods, the reader is directed to the works of Heuer, Heuer and Pherson, Clark, and the Center for the Study of Intelligence.

improve analysis must be based on a detailed cognitive understanding of analysts' methods and thought processes. These reports, however, received little attention at the time they were published. It was only decades later that intelligence professionals began to describe analysis as a "cognitive process." Rob Johnston, formerly with the CIA's Center for the Study of Intelligence, referred to intelligence analysis as "the application of individual and collective cognitive methods to weigh data and test hypotheses in a secret socio-cultural context" [4, p. 4]. Former NSA Technical Director David Moore described how intelligence assessments require "the core cognitive skills of critical thinking: interpretation, analysis, evaluation, inference, explanation and self-regulation" [5, p. 59].

Currently, the U.S. Office of Personnel Management officially describes intelligence research and analysis as a continuous process of evaluating information collected from a variety of sources, drawing inferences from its analysis and interpreting those inferences in the context of national security policy [8].

9.2.2 Assessing Analytic Quality

Because intelligence analysis is not an exact science, judgment or decision quality is a difficult metric to measure. In fact, more has been written about what *not* to measure, or how to assess failure in hindsight, than what constitutes a reasonable measure of analytic quality [9]. Often, the value of an intelligence product is considered from the point of view of the consumer. Policy guidelines prescribe vague standards in terms of timeliness, availability and usability—all extrinsic attributes that evaluate intelligence based on how well the reports are received by decision makers. In addition, both the Department of Defense[2] and the Office of the Director of National Intelligence[3] justifiably include accuracy and rationality criteria in their quality standards. Further guidance requires analysts to "identify and explain the strengths and weaknesses of alternative hypotheses, viewpoints, or outcomes in light of both available information and information gaps" (pp. 3–4).

Analysis in this domain is heavily impacted by uncertainty and an analyst rarely, if ever, gets a complete picture of what is actually happening. As such, most judgments result from a combination of observations, assumptions and inferences *at the time of the analysis*. Situations can change rapidly and dramatically, and problems are of a non-deterministic nature—i.e. analysts must consider multiple plausible futures.

---

[2]The DoD policy states that "intelligence must be factually correct, relay the situation as it actually exists, and provide an understanding of the operational environment based on the rational judgment of available information" [10].

[3]ODNI standards require that analysis "accurately characterize the information in the underlying sources and explain which information proved key to analytic judgments and why" and take into account such "factors significantly affecting the weighting that the analysis gives to available, relevant information, such as denial and deception, source access, source motivations and bias, or age and continued currency of information, or other factors affecting the quality and potential reliability of the information" [11].

Due to these and other factors, accuracy turns out to be a poor objective criterion for assessing analytic quality.

The rationality criterion, on the other hand, offers greater potential as a quality indicator. In applying this principle, two primary considerations are made: the quality of the information analyzed and the thoroughness of the analytic process followed. A number of information quality attributes are mentioned above, e.g. source credibility, information currency, the potential for deception, etc. Others may include redundancy, diagnosticity and precision. The objective in the purposeful evaluation of data quality attributes is to constrain the decision space to the data points necessary for a sound judgment.

A rigorous analytic process, being necessary for high-quality analysis, is the second component of the rationality criterion. Here, tradecraft and methodology come into play as the analyst must apply logical argumentation and point out the strengths and weaknesses of alternative hypotheses, identify information gaps, identify less probable but highly impactful outcomes, etc. [11]. The challenge is to make the entire process transparent so as to facilitate a proper evaluation.

9.2.3 Judging "Correctness"

Many intelligence problems are judgmental in nature, meaning that they have no objectively correct solution. The Intelligence Community has struggled with developing objective intrinsic quality measures that provide a reliable and consistent mechanism for evaluating analytic judgments and how they are made. Accuracy in intelligence products should be an obvious goal, but in practice, the implementation of a quality metric based on accuracy has not met with much success. Complete objectivity in this discipline may be impossible to achieve and it seems that every proposed quality assessment measure reveals some amount of inherent subjectivity. However, by externalizing the analytic process—i.e. exposing the steps that an analyst follows and making judgments and decisions more transparent—the effects of debiasing manipulations on the analytic process will become observable. Fortunately, prior research on judgmental decision-making has established evaluation protocols for use in experimental settings.

Judgment and decision-making research comparing individual to group performance on judgmental tasks frequently makes use of simulations or game-based tasks. "Correctness," or response quality, is determined by comparing subject performance to that of domain experts. For example, Littlepage, Robison [12], Fiedler [13] and others used a simulation game called Desert Survival, which situates subjects in the desert following the crash of their small plane and requires them to salvage and rank the items most important to their survival. Responses are scored according to an answer key that was developed by survival experts.

9.3 Impact of Heuristics and Biases on Intelligence Analysis

Tversky and Kahneman [14] provided the most widely adopted explanation for reduced performance on cognitive decision tasks—the reliance on mental shortcuts, or *heuristics*, in making judgments. Since then, decades of research have produced scores of findings about dozens of different types of cognitive biases in human judgment and decision making.

A *heuristic* is a mental technique that allows an individual to reduce a complex problem to a simpler one—basically, a form of abbreviated thinking. Heuristics, in this view, can be thought of as deviations from rational or normative decision processes. Although heuristics serve a functional need to preserve cognitive capacity and are often beneficial, they can lead to systematic errors or *biases* when problems are oversimplified [15]. By most accounts, errors generally result from the application of heuristics rather than insufficient knowledge. Experts and novices exhibit similar biases [16, 17] and biased judgment does not appear to be tied to cognitive ability [18, 19].

Cognitive biases are believed to have a strong impact on analytic performance, particularly when a situation calls for evaluating incomplete or uncertain information or when analytic judgments must be made within severe time constraints [20]. Even though professionals in fields such as medicine and law enforcement, are highly trained, neither is immune to the effects of judgment bias. Determining what information is germane to an intelligence problem, weighting information appropriately and thoroughly, and objectively evaluating all available evidence are critical steps in intelligence analysis. Heuristics are widely applied during these activities and this section will examine a few of the most common biases that may result and their possible impact on analysis.

There are hundreds of labeled cognitive biases documented in the scientific literature on this topic [21]. Many of them, because they impact problem solving and judgment making, are believed to influence the quality of intelligence products and the analyst's confidence in his or her accuracy. Despite decades of research on identifying and cataloging biases, relatively little work has been done to overcome the negative effects they have on intelligence analysis. Section 9.4 describes the progress to date on bias mitigation.

9.3.1 Confirmation Bias

Also referred to as "biased assimilation," the *confirmation bias* is one of the earliest documented and longest-studied sources of judgment error. Appearing in the prejudice literature of the 1950s and social behavior literature of the 1960s and 1970s, confirmation bias is one of the strongest and most pervasive of all cognitive biases. Simply put, confirmation bias is an information and memory search strategy driven by prior beliefs. Information that confirms one's prior beliefs is likely to be accepted at face value while disconfirmatory information is apt to be received with skepticism or rejected outright [22]. Confirmation bias can sometimes result in premature closure; i.e. terminating the analysis or ending the search for relevant information too early and forming a conclusion based on insufficient evidence.

A confirmation bias in an ambiguous intelligence problem can "compromise objectivity through neglect of conflicting evidence and judgments not reflective of the entire evidence spectrum" [23]. The effect can be exacerbated when an individual becomes psychologically invested in a particular. Individuals impacted by this effect exhibit biased recall and interpretation of favorable facts as well as biased judgments influenced by unbalanced weighting of those facts. An important implication of the confirmation bias is that once evidence has been processed and judged to be confirmatory or disconfirmatory, it continues to support the prior belief even when that belief becomes subject to new disconfirmation or the original evidence itself is later disconfirmed [24]. This is what Heuer [20, p. 124] refers to as the "persistence of impressions based on discredited evidence".

9.3.2 Illusory Correlation

Illusory correlation, as the name implies, is a systematic error in perceiving an expected relationship between events, people, groups, behaviors, etc. when no such relationship actually exists [25]. Heuer offers a more precise definition: "the extent to which the intuitive perception of covariation deviates from the statistical measurement of covariation" [20, p. 140]. Social psychologists have studied illusory correlation in the context of social stereotypes, attempting to pinpoint the origin of the correlation by teasing apart factors such as memory effects, response bias and information source. In the latter, the bias results from an early assessment of the source of the information rather than the information itself [26].

Heuer offers several examples of assumptions that analysts and others often make, but which are typically based on intuition rather than statistical analysis [20]. Correlations between worsening economic conditions and increased political opposition support, or between military rule and the demise of democracy, may be accurate in many cases but are usually unsupported by evidence and analysis. Other more flagrant examples are the correlation of strange behaviors to full moons, college dropouts to business successes, or acts of violence to particular ethnicities or faiths.

9.3.3 Absence of Evidence

As an old axiom states, "the absence of evidence is not evidence of absence." Because key information is usually lacking in intelligence analysis, accurate judgments cannot be made based only on consideration of available data. An analyst must be able to recognize what relevant information is lacking and factor it into their judgment.

Furthermore, the lack of key pieces of data should, normatively speaking, result in lower confidence assessments and less extreme judgments.

Studies by Garcia-Retamero and Rieskamp [27] showed wide disparity in how people treat missing information. Pieces known to be missing were sometimes treated as negative, positive, or "average" as compared to other information, or were ignored entirely. However, perceivers did not always recognize the absence of relevant information in forming judgments. Individuals having more knowledge about the tested subject tended to be more sensitive to relevant omissions [28]. Kardes and Posavac [29] found that individuals who were insensitive to relevant missing information tended to make extreme judgments, whereas individuals sensitive to important missing information developed more moderate and appropriate assessments.

9.3.4 Irrelevant Evidence

Research has demonstrated that certain judgments based on a mix of relevant and irrelevant information tended to be less extreme than judgments based only on the relevant information [30]. Labeled the "dilution effect", the phenomenon was reliably observed when a mix of information was considered in the rendering of the initial judgment. The effect disappeared when the initial judgment was made based only on relevant information and later reconsidered in view of irrelevant data [31].

A similar issue is the effect that irrelevant hypotheses have on judgments. Most studies that have addressed how people analyze alternatives have assumed that an individual will develop a focal hypothesis (usually the first one considered) along with one or more alternative hypotheses [32]. Alternative hypotheses are considered relevant if they have some possibility of occurring, whereas irrelevant hypotheses are essentially impossible [33].

Studies have shown that the perceived likelihood of a focal outcome tends to be disproportionately influenced by two things: the strength of the strongest alternative hypothesis [34] and the distribution of alternative outcomes [33]. For example, assume that an analyst has a moderate amount of information that supports Hypothesis A and a moderate amount of conflicting information that supports alternative hypotheses, B to F. Hypothesis A will seem much more likely if the conflicting information is distributed among the alternatives, B to F, in small amounts than if there is a moderate amount of support for a single alternative, say Hypothesis B.

9.3.5 Overconfidence Bias

It is a common finding in the cognitive psychology literature that individuals tend to display systematic errors in judgment when asked to assess the accuracy or correctness of their own performance on intellectual problems [35]. In particular, self-reported assessment of accuracy on cognitive tasks tends toward overconfidence, while self-reported assessment of visual perceptual judgments often produces under-confidence [36]. Experimental findings on a variety of tasks suggest that individual differences play a role in the degree to which confidence tends to be over-estimated or under-estimated [35, 37].

Overconfidence can result in serious consequences in intelligence analysis. In fact, overstated confidence may have resulted in a decade-long war that ultimately cost a CIA Director his job.[4] In the aftermath of the flawed intelligence assessment of Iraqi weapons of mass destruction, Congress and the President enacted Intelligence Community reforms that, amongst other things, called for analytic confidence to be expressed in finished intelligence products. However, as Peterson [39] points out in his thesis, no standardized method of assessing or communicating analytic confidence has yet been published and few studies have examined confidence in realistic settings. In one recent study that investigated judgment accuracy and confidence in an intelligence context, an inverse relationship between accuracy and confidence was found [40]—higher self-reported confidence ratings were associated with incorrect responses.

9.4   Mitigating Bias Effects

Despite decades of research, there is still no clear and comprehensive formula for avoiding cognitive biases in complex analysis and decision making. Though heuristics and biases have been widely studied, much of that research has been devoted to identifying and labeling biases rather than mitigating their effects [41]. Furthermore, bias mitigation studies are typically conducted using simple, tightly-controlled laboratory tasks rather than complex, realistic problems. Intelligence specialists and practitioners have proposed and instituted numerous tools and techniques designed to reduce bias and improve critical thinking (e.g. "structured analytic techniques" [42], "analysis of competing hypotheses" [43], etc.) but little has been done to exper-imentally validate their impact on analytic quality.

Research suggests that there are limits to what can reasonably be expected of individuals in terms of applying bias mitigation techniques. Certain styles of intervention are easier and more natural than others, but some biases have systemic roots that cannot be easily accessed using individual mitigation strategies. Moreover, biases can be difficult to defeat in individuals who fail to acknowledge their existence (i.e. the bias blind spot [16]). Yet, a handful of studies suggests that debiasing analysis is not only theoretically plausible, but that bias mitigating techniques can be tested and shown to be effective in mitigating many of the most common cognitive errors committed in the analysis of intelligence information. Furthermore, many of these techniques will

---

[4]It was reported by investigative journalist Bob Woodward [38] that then CIA Director George Tenet personally supported the Iraq weapons of mass destruction assessment, assuring President Bush of its accuracy and advising him to tell the American people that it was a "slam dunk case" in order to gain popular support for military operations.

be ergonomically practical[5] and ecologically valid—two important considerations for this critical profession.

Past research on attempts at mitigation can be generally divided into training interventions and procedural interventions. Training interventions take place in a classroom or similar setting outside the workplace while not performing work functions. The goal is to educate personnel about the existence and effects of heuristics and biases. Conversely, procedural interventions are designed to be integrated into operational work processes, thus mitigating bias effects whilst working. Under these conditions, workers need never be made aware of the operation of cognitive heuristics.

9.4.1 Training Interventions

The most basic form of bias mitigation reported in research, is teaching individuals how to avoid succumbing to heuristics under circumstances when bias is likely to result. Fischhoff [45] initially approached the issue simply and directly: he taught subjects about how a heuristic produces a bias and then instructed them not to apply the heuristic. Unfortunately, most research shows that general approaches to critical thinking instruction rarely generalized beyond the domain or the tasks on which they were taught [46, 47] and showed mixed results, at best [48]. One explanation for the lack of success using these sorts of techniques is that many heuristics are likely applied subconsciously and, therefore, difficult to recognize and control [49–51]. Another stated that to be effective in mitigating cognitive biases, the mitigation strategy must be closely related to the circumstances under which the bias is produced [49].

For an example of within-domain research, we look to the literature on studies directly applied to intelligence analysis. One recent research program conducted by the Intelligence Advanced Research Projects Activity (IARPA) funded research teams to "create Serious Games to train participants and measure their proficiency in recognizing and mitigating the cognitive biases that commonly affect all types of intelligence analysis" [52]. Serious games are defined as video games designed for training or education rather than purely for entertainment [53]. Teams studied two fundamental questions: whether serious games were effective in increasing knowledge of biased decisions among participants and whether serious games were more effective than instructional videos. Results confirmed both hypotheses—that serious games were more effective than educational videos and that subjects who played the games outperformed those who did not in recognizing and mitigating biases.

---

[5]Cognitive ergonomics is defined by the International Ergonomics Association, the concern for "mental processes, such as perception, memory, reasoning, and motor response, as they affect interactions among humans and other elements of a system. The relevant topics include mental workload, decision-making, skilled performance, human-computer interaction, human reliability, work stress and training as these may relate to human-system design" [44]. Cognitive ergonomics addresses cognitive well-being and performance in work settings and operational environments.

It should be noted however that participants were aware that the study was about bias and bias was made salient by the experimenter. None of the studies examined the change in bias manifested when the concept of bias was not intentionally made salient.

9.4.2 Procedural Interventions

Early attempts to showcase bias mitigation instead demonstrated that biases are highly robust and tend to be resistant to correction [20, 54]. Many early documented interventions sought to improve the estimation of outcome likelihood by instructing subjects how to calculate probabilities or construct Bayes models as part of their work. Even though these techniques incorporated tabular or graphical tools, most were largely ineffective at reducing biased judgments. Later attempts focused on procedural mitigation strategies to elicit greater objectivity when evaluating evidence and considering possible causes or explanations [55, 56]. The emphasis was on developing bias-reducing methods that could be used on the job without making heuristics or biases salient. Some of these techniques showed promise, particularly when presented using visual tools.

Analyzing alternative hypotheses

"The evidence seems … fairly compelling that people do not naturally adopt a falsifying strategy of hypothesis testing. Our natural tendency seems to be to look for evidence that is directly supportive of hypotheses we favor and even, in some instances, of those we are entertaining but about which we are indifferent" [57, p. 211]. Graber and Franklin [58] reported that cognitive errors in medical diagnosis were most often due to faulty hypothesis generation and evaluation and less often due to insufficient knowledge or information gathering. The technique of analyzing competing hypotheses attempts to circumvent our natural tendencies by explicitly identifying all reasonable explanations or conclusions and comparing them to one another to determine which is most likely the correct one [20].

Heuer developed a formal technique termed the Analysis of Competing Hypotheses (ACH) to help intelligence analysts overcome some of their cognitive limitations. ACH is an eight-step procedure "grounded in basic insights from cognitive psychology, decision analysis and the scientific method" [20, p. 95] and designed to help analysts avoid common mental pitfalls [42]. Though it has received little scientific investigation in actual work settings, this technique is now taught in training courses throughout the Intelligence Community. ACH attempts to simplify the problem of evaluating multiple explanations simultaneously.

There is a limited amount of empirical laboratory support for the ACH-style approach. Various studies have shown that the development of alternate hypotheses (also referred to as counterfactual primes) and "consider the opposite" techniques heightened awareness and focus of relevant alternatives [59], debiased primacy and recency effects [60], increased confidence in judgments [61] and helped subjects avoid the "first instinct fallacy" (i.e. the tendency to believe that the one's first answer is always the correct or best answer) [62]. Furthermore, confidence inquiries were shown to encourage participants to increase their consideration of alternatives [63].

Other research has been less supportive, however. In the only known experimental study of ACH on a realistic intelligence problem, Folker [64] found a statistically significant improvement in the quality of analytic judgments only on the simpler of two problems. Encouraging individuals to consider multiple alternatives only helped when the generation of alternatives was done easily [59] and the number of alternatives was small [58, 65]. Other research demonstrated an individual's preference for considering evidence against only a single focal hypothesis [66–68]. A post-9/11 ethnographic study [4] found that organizational cultural norms—in particular, a guild-like tradecraft culture coupled with a strong production orientation—limited the use of formal analytic techniques, which led former National Intelligence Council Chairman Greg Treverton [69, p. xi] to conclude that ACH, in particular, "isn't all that valuable" for journeyman or expert analysts. While it seems that available evidence offers hope for improving analytic quality by evaluating alternative hypotheses, perhaps a more light-weight, less time-consuming way of performing the necessary steps might offer a more practical solution.

External review

Research shows that individuals sometimes perform better when they know that their work will be subjected to external review or critique [70]. Markman and Hirt [71], in the context of allegiance bias, showed that when a strongly biased individual was told to anticipate a discussion with another individual of unknown allegiance, predictions were less biased. Cook and Smallman [23], however, failed to achieve a significant reduction in bias by exposing subjects' work to other individuals for review. The authors offered some possible reasons for the lack of significance: firstly, the lack of social pressure, since the performer and the reviewer did not collaborate during the task, and secondly, the lack of awareness of each other's analytic "credentials." Because of the lack of experimental investigation on the impact of external factors on individual analytic judgments, the effectiveness of external review remains an open question.

Checklists

Checklisting is used by professionals in a number of demanding occupations such as airline pilots, maritime crews and nuclear plant operators in order protect against foreseeable but avoidable errors. The basic idea behind checklists is to provide an alternative to reliance on intuition and memory in problem solving and thereby resist the effects of biases and failed heuristics. While checklists often contain information that is fairly obvious to trained personnel, important steps are often forgotten because of distractions, stress or time pressure.

Ely and Graber [72] proposed three categories of checklists for use in diagnosis. The categories correspond to the levels of Croskerry's [73] cognitive forcing strategies described in the next section. At the most abstract level, a universal checklist helps physicians to understand and apply basic cognitive skills. Next, a general checklist guides practitioners through a differential diagnosis while attempting to avoid common cognitive pitfalls. A final set of checklists addresses potential sources of error in order to improve diagnosis of specific diseases. Graber [74] suggested that medical students become accustomed to using a simple checklist for the diagnostic process and Heuer [20] proposed a process checklist for intelligence consisting of six key steps: problem definition, hypothesis generation, information collection, hypothesis evaluation, hypothesis selection and the continued monitoring of incoming information.

A checklist seems, at face value, to be a reasonable mechanism to reduce the frequency and magnitude of process errors. The use of checklists has been investigated in medical settings; for example, Haynes, Weiser [75] implemented a 19-item surgical safety checklist designed to improve team communication and consistency of care. However, there has been no reported evidence to support the effectiveness of checklists in reducing cognitive errors in actual settings for either medicine or intelligence.

Cognitive forcing strategies (CFS)

Croskerry [73] coined the term to describe three levels of metacognitive strategies to compensate for latent judgment errors: universal, generic and specific. Universal strategies entail acquiring general knowledge about heuristics and biases, while generic and specific strategies acquire knowledge about heuristics common to the medical decision domain and to specific situations and scenarios, respectively. Essentially, CFS is a general taxonomy for organizing any sort of bias mitigation technique, to include the others described previously. In support of this idea, Trowbridge [76] offered twelve tips for avoiding diagnostic errors. The tips present universal and general guidance but offer no evidence to support their effectiveness. Sherbino and Dore [77] performed one of the few known studies to evaluate the effectiveness of such strategies and found training retention was poor and difficulty in applying the techniques was high.

Establish judgment criteria

Kardes and Posavac [29] conducted experiments to investigate the effectiveness of two procedures for improving judgment by increasing sensitivity to missing information. When subjects were insensitive to important missing information, overly extreme evaluations were formed. However, when they became sensitized to the missing data, they arrived at more moderate and appropriate conclusions. Sensitivity to missing information was increased by encouraging individuals to consider their criteria for judgment before receiving the information, and by asking them to rate present and missing attributes before providing their overall evaluations. Both procedures were effective for improving judgment by reducing what the authors called "omission neglect."

Heuer [20] expanded the Kardes procedures by suggesting that analysts should identify explicitly those relevant variables on which information is lacking, consider alternative hypotheses concerning the status of these variables and then modify their judgment and adjust their confidence accordingly. Furthermore, they should consider whether the absence of important information is normal or is itself an indicator of unusual activity or inactivity. Like the others described, experimental support for this technique is lacking; it seems reasonable to investigate, however, whether expressing the relevant variables or judgment criteria a priori would reduce the impact that missing or irrelevant information bears on analytic judgments.

Disfluency

Fluency can be defined as the ease with which cognitive processing occurs when completing a mental task. Disfluency, then, refers to the interruption of the smooth flow of thought. Studies have reported that subtle, sometimes unnoticed, changes in stimuli, such as using a less common font, can induce a deeper, more effortful cognitive process and result in increased reflection and greater objectivity during a problem-solving activity [78]. Pursuing this idea, Kretz [79] designed an experiment based on this principle of disfluency, exploring the effects of a variety of minimally and mildly invasive bias mitigation techniques on analytic judgments.

When considering how to study procedural interventions for bias mitigation in intelligence analysis, using ergonomically practical and ecologically valid methods, the literature makes several additional points. Firstly, the mitigation strategy should be closely related to the circumstances under which the bias is produced [49]. Secondly, when designing a mitigation strategy, biases should not be treated in isolation as many heuristics ultimately produce the same biased effects in judgments. So, rather than dealing with specific biases, the focus should be on improving the process and performance of analysts in terms of outcome measures, such as analytic quality. Finally, changing how humans are wired to think is a difficult task. Changing the environment in which they think is much easier and can improve outcomes just as effectively [80].

9.5 Experimental Methods

The preceding sections made a convincing case that cognitive heuristics often produce biased thinking that could lead even trained experts like intelligence analysts to form biased judgments. That phenomenon, coupled with the disdain many expert analysts have for cumbersome, formal bias-reducing methods, is cause for concern among senior intelligence officials. The overarching goal of research is to investigate techniques to enable analysts to avoid the sorts of heuristics that lead to poor quality judgments without creating time-consuming process impediments. It is important for us to develop a broadly applicable methodology that satisfies the requirements of rigorous experimental design while recognizing the need to evaluate analytic quality in complex, realistic problems.

There are many quasi-experimental and usability testing alternatives available to consider when experimental conditions cannot be met. However, those are outside the scope of this particular chapter. To address the need described above, this chapter outlines a general experimental framework for investigating bias mitigations for intelligence analysis.

This section will begin by offering a set of considerations germane to this effort followed by a short description of relevant past work with emphasis on the designs. Finally, we distill the information into a set of guidelines on how to conduct an experimentally rigorous investigation of bias mitigation techniques.

9.5.1 Considerations

Realism. Perhaps the most distinguishing characteristic of experiments of this type is the one posing the greatest challenge [81]—the extent to which they emulate real world conditions. As an applied research problem, the goal of a mitigation technique is to achieve an improvement in actual job performance. External validity is an important consideration, for without it, incorrect conclusions may be drawn about the efficacy of the technique. Several realism factors are important in this type of research—experimental realism and mundane realism. The former reflects the degree to which subjects are able to treat the experience as realistic, whilst the latter refers to the extent to which the experimental situation is similar to situations people are likely to encounter outside of the laboratory [82]. Another related concept is that of ecological validity, the degree to which the experiment is representative of a real-life setting [83].

Subjectivity. Problems and tasks can be categorized in many ways. One such continuum describes tasks in terms of subjectivity. On one extreme, intellective tasks have a demonstrably correct answer that is universally agreed upon (e.g. math problems, on the other extreme, judgmental problems have no such answer and are entirely subjective (e.g. the quality of a work of art). Many problems, particularly complex ones, fall somewhere in between these two extremes. Intelligence problems, because they contain elements of uncertainty (deception, ambiguity, incompleteness, etc.), tend toward the judgmental end of the spectrum. The implication, as described earlier in the chapter, is that it is more difficult to judge the correctness of subject responses in an experimental task.

Iteration. One-shot games are used extensively in behavioral economic research, where the participant receives a set of data or facts and is asked to render a single judgment. The Prisoner's Dilemma[6] is a well-known example of such a game. One-shot games are limited in that they cannot examine the effect of changes such as discredited sources or contradictory information. In contrast, *multi-shot games* introduce multiple sets of information in phases. In this format, individuals are required

---

[6]The Prisoner's Dilemma is a two-player game that shows why two completely rational individuals might choose not to cooperate, even if it appears to be in their best interests to do so [84].

to re-examine their judgments and possibly update their beliefs as new information is received. This type of task is much more representative of real-world problems of an intelligence nature.

9.5.2 Prior Studies

Only a handful of studies have examined bias mitigation in the specific context of intelligence analysis problems in experimentally rigorous ways. Because the number of studies is small, we can summarize them here. Readers are encouraged to read each of these publications in their entirety if contemplating designs of their own. Three relevant studies are discussed.

Study 1: Folker (2000)

Summary: In the first study of its kind, Folker [64], experimentally compared the analytical conclusions drawn by two sets of trained analysts who solved the same decision tasks. Individual analysts in the control group were not told to apply any specific techniques and received no training (i.e. "intuitive" analysis). Analysts in the experimental group were trained to use a specific structured methodology, the matrix-based analysis of competing hypotheses technique, to aid them in their analysis. The task consisted of military scenarios with a map and evidence presented as text. Subjects read the evidence, studied the map and rendered predictions. The predictions were scored as either correct or incorrect based on the opinions of expert judges. The scores were then compared statistically to determine whether the experimental group did significantly better than the control group. Fisher's Exact Probability Test was used as a test of statistical significance because of the small sample size. A pretest questionnaire consisting of demographic and prior experience questions was administered to all subjects. Folker reported that the experimental group outperformed the control group in only one of the two scenarios.

Though the map was not part of the manipulation, the organization of evidence and hypotheses into a matrix constitutes a visual analytic technique and hence was the bias mitigator being tested. Though presented as an intellective decision problem, Folker acknowledged that intelligence scenarios are by their nature more judgmental. Folker described the use of a pilot study to validate and refine his design. This study made use of open ended questions, which are harder to score but offer insight into the subjects' thought process. The task was time-bounded and although each participant completed more than one scenario, the study was a between-subjects design since there were two groups under different conditions being compared.

Study 2: Cook and Smallman (2008)

Summary: In this study, the authors tested two debiasing interventions intended to help analysts objectively weigh the full spectrum of evidence and its relationship to hypotheses [23]. The interventions included: (i) a graphical layout for integrating evidence and (ii) seeing other analysts' evidence assessments. Each of the two interventions was tested under two conditions—graphical versus text or seeing others versus own assessments only—making for four experimental conditions in a repeated-measures design (i.e. no control group). Four fictitious vignettes were used, each containing a hypothesis, key background information and relevant evidence that either supported or contradicted the hypothesis. Correctness was determined by comparison with responses from a panel of expert judges.

The exercise sessions were conducted on a computer using a collaborative decision support tool called JIGSAW.[7] Participants included both experts and novices who assessed evidence, related evidence to hypotheses and ranked evidence in order of importance but did not "solve" an investigative problem. A 2×2 repeated measures ANOVA was used to compare results and test for significance. Cook and Smallman found that the graphical evidence layout promoted more balanced and less biased evidence selection than a text-based presentation. However, seeing other analysts' assessments did not produce a significant improvement.

Study 3: Kretz (2015)

Summary: This most recent study tested the principle of disfluency and compared four debiasing interventions against a control (intuitive) group [79]. Participants in each of the four experimental groups received instruction in one of the visual bias mitigators: (i) check the box after reading each piece of evidence; (ii) read a list of possible hypotheses before solving the problem; (iii) make a list of possible hypotheses; and (iv) map each piece of evidence to the hypotheses to which it relates. The task was described as an Analytic Decision Task (ADT), a vignette-based multi-shot game—a realistic form of complex judgmental task consisting of a three-part scenario that required participants to render judgments at multiple times throughout the task.

Responses were open-ended and were coded by a team of psychology research assistants. Correctness was scored by comparing subject results to scores obtained from an expert panel of intelligence experts. Subjects also reported their confidence in each of their answers and completed a sequence of cognitive tests as well as a demographic questionnaire. The design was improved and refined during pre-trial pilot sessions. Data was analyzed using a single factor, between-subjects ANOVA experimental design. When comparing quality in terms of final responses, the differences between mitigators and the control were not significant. But when comparing improvement in terms of providing better answers as the game progressed, significant differences were seen. None of the cognitive test findings were found to be correlated with quality or improvement, but weak correlations between cognitive test scores and self-reported confidence were significant. The study also reported a statistic labeled *response inertia* (RI), which indicated a subject's apparent unwillingness to change their answer in light of further evidence. Interestingly, RI was shown to be significant.

---

[7]Joint Intelligence Graphical Situation Awareness Web; see [85].

9.5.3   Experimental Design Framework

Drawing from the useful features of the studies reported above, a reference framework was developed for the experimental evaluation of bias mitigations applied to problems of an intelligence nature. Several options are presented throughout, in order to cater to different needs. The discussion below presumes at least a basic understanding of experimental research.

At this point, several assumptions are made. These steps are germane to any experimental study. First, the technique(s) or intervention(s) to be tested has been chosen. Second, the research questions have been developed and stated. Third, informed consent to participate has been obtained from all subjects.

The remainder of the chapter discusses the nuts and bolts of the study design as it will apply to this sort of problem.

- Determine the variables
- Design the task
- Sample and assign to groups
- Determine how the data will be analyzed
- Pilot and refine.

Determine the variables

The experimental variables are derived from the research questions and hypotheses. The independent variable (IV) is the condition being manipulated by the experimenter. In this case, it is the bias mitigation technique. This IV may take on multiple values if more than one technique is being tested. In Study 3, four separate disfluency techniques were tested against the intuitive analysis condition (i.e. the control group).

If the mitigation technique(s) will be tested under more than one condition, adding additional IVs is appropriate. If, for example, you want to study the mitigation techniques under both bounded and unbounded conditions with respect to time, then a time bounding IV would be added with two conditions. Other possible IVs for a study of this type include individual versus collaborative (as in Study 2) or graphical versus text (also in Study 2).

It is important to emphasize that IVs in experimental design are manipulated variables, suggesting that each participant has an equal chance of being assigned to each of the conditions. Random assignment to conditions is a requirement in experimental design. Variables such as age, gender or experience are naturally occurring and non-manipulated, making them suitable only for quasi-experimental studies.

The dependent variable(s) (DV), in contrast to being controlled by the experimenter, are measured by the experimenter—they change in response to changes in the IV (i.e. they depend on the value of the IV). The DV used in Studies 1, 2 and 3 was a measure of analytic quality as reflected by the "correctness" of subject responses. Quality in these studies was seen as the inverse of bias—high quality responses were judged to be less biased and vice versa.

Operationalizing the DV is straightforward for intellective problems. Because they have demonstrably right or wrong answers, intellective problems can easily be scored as right or wrong by the experimenter. Problems of a more judgmental nature are more complex and require the validation of responses; i.e. determining the "best" response, or ranking possible responses in terms of quality. Empaneling one or more outside experts is a common approach for performing such a validation. Study 3 describes in detail how expert analysts were recruited and how their independent analyses were used to develop a scoring standard. This standard consisted of list of foreseeable responses ranked in terms of how well the evidence supported them. In an airline crash scenario, for example, possible explanations for the crash were poor weather, pilot error, system or part malfunction, or an intentional act of terrorism. These explanations were ranked in terms of how likely they were based on the evidence presented and the ranks were then used to evaluate subject responses.

Yet other DVs may be quantitative in nature, which simplifies the process of operationalizing them. Confidence self-reported on a scale of 0 to 10, cognitive test scores and rankings of evidence relevance are examples, from Studies 2 and 3, that fit this description.

Sampling and assignment to groups

When it comes to choosing participants, recruiting and selecting individuals who resemble analysts in terms of skills and experience offers the greatest external validity for the study. The actual population of analysts is small and unlikely to be accessible, but the experimenter may have access to former analysts or professionals in occupations with similar characteristics (e.g. engineers, scientists, financial analysts, business analysts etc.). If sampling from a university student body or the general population, the research findings are valuable but the experimenter must make a stronger case to support any claims of generalizability or predictability.

The number of subjects to recruit depends on the number of groups and the desired statistical power of the study. The experimenter should perform a power analysis in order to determine the appropriate sample size for the study. This is an important step in order to establish statistical significance; i.e. a valid claim that the intervention resulted in the observed outcome. A power analysis will report the probability of detecting an effect[8] with some degree of confidence.[9]

Each participant will be assigned to a group. The number of groups to be tested depends on the number of IVs chosen and the number of conditions associated with each IV. It also depends on the type of measure being used: independent or repeated measure.

If the experiment will investigate whether one or more mitigation techniques improve quality, then an independent measures/between-subjects design is sufficient. This design involves a single IV to compare the performance of separate groups of subjects against a control group. Thus, no subject participates in more than one group.

---

[8]The term effect is used to describe a change in a dependent variable that is presumably caused by the manipulation of an independent variable.

[9]For a primer on conducting power analyses, see [86].

This is the design followed by Studies 1 and 3. As mentioned earlier, the random assignment to conditions is a necessary step in experimental studies, so subjects must be assigned to a group without regard for age, gender, experience or when they signed up.

If the experimenter believes there could be a confounding factor,[10] then the experimenter should consider using a matched pair or randomized block technique[11] to eliminate the confound. For example, subjects who have analysis in their work history would reasonably be expected to perform better than subjects who lack such experience. Pooling homogenous groups with a randomized block technique or dividing subjects equally, *but randomly*, between the groups using a matched pair technique would eliminate the effect of analytic work experience on the outcome.

If the experimenter's access to participants is limited or the sample size is small, then a repeated measures design should be considered. Here, only one group of participants is exposed to each of the conditions. Study 2 makes use of this design. The main concern with repeated measures studies is order effects; i.e. that some subjects will do better in second and subsequent tasks because they have practiced the task, or that some of the subjects will do worse in later tasks due to mental fatigue. Counterbalancing the tasks between participants can alleviate many of the order effects.

Design the task

Consistent with the realism discussion above, the task should have the same qualities as a typical intelligence problem: ambiguity, uncertainty, irrelevant data, discredited sources, changing situations, etc. Consider a *complex judgmental task*. This type of task has two key characteristics. As a complex task, it is an amalgamation of smaller subtasks; in this case, generating hypotheses, weighing evidence and determining the most likely hypothesis. As a judgmental task, it has no demonstrably "right" or "wrong" answer. Because many intelligence problems share these characteristics, the complex judgmental task is a suitable mechanism for use in intelligence studies.

The choice of a one- or multi-shot format depends on the nature of the bias being addressed. If the study will require participants to render a single judgment based on a simple scenario with limited data, a one-shot game like those in Studies 1 and 2 is sufficient. If the study will investigate participant judgments over time as situations change and evolve, or as information is presented and later discredited or contradicted, then a multi-shot game such as the Analytic Decision Task in Study 3 offers greater options.

---

[10]A confounding factor is any factor other than the independent variable that affects the result; i.e. accounts for some of the variance in the dependent variable.

[11]In a randomized block design, the experimenter divides the sample into homogenous blocks (e.g. based on years of experience: 0–2 years, 3–5 years, 6–10 years, etc.) if the variance within each group is likely to be less than the variance between them. In other words, if subjects in the 0–2 group are expected to be relatively similar, but different overall from the 3–5 and 6–10 groups, then blocking would offer better estimates of effect size. Caution: if that assumption turns out to be incorrect, then blocking may offer worse estimates.

For the higher validity, the scenarios used in Studies 1–3 were constructed from actual events or situations using openly available information, though names and places were altered to reduce the likelihood of remembrance. The key criteria for inclusion were: (i) the information must suggest multiple plausible hypotheses with no undisputed outcome, and (ii) the information must contain important elements of intelligence problems such as discredited evidence, ambiguous statements, incomplete data, and irrelevant and redundant facts.

Finally, because the task is complex and variance in the outcome can come from individual difference factors unrelated to the independent variable, collecting additional data is recommended. For example, Studies 1 and 3 collected demographic information, and Study 3 administered a series of cognitive tests to assess participants' cognitive style and cognitive reflection. This additional data can be used to test for correlations between performance and demographic or cognitive factors in order to explain the observed variance in the dependent variable.

Determine how to analyze the data

The primary question to answer is whether or not the differences in the participants' judgments were caused by the use of the bias mitigation technique. There are several methods available to test that hypothesis and the choice of which to use depends on the size of the sample and the number of populations (groups) being tested.[12]

The most commonly used tests in experiments of the type discussed here are the t-test and the analysis of variance (ANOVA). These are both parametric tests, which make certain assumptions about the type and distribution of the data. Non-parametric tests make no such assumptions and may be a good alternative when the guidelines for parametric analysis cannot be met.

Generally speaking, parametric tests perform best for relatively large samples (group size > 15). The t-test is designed for use when comparing two populations; e.g. a control group and an experimental group. It is appropriate for comparing means of both independent (unpaired) and paired samples. The ANOVA is used when more than two populations are being compared. In the case of a single IV with three or more conditions, a one-way ANOVA is appropriate. Study 3 uses a one-way ANOVA to compare five group means under a single IV. If the design of the experiment includes two or more IVs, a factorial ANOVA must be used. In Study 2, experimenters designed their study to include two IVs each with two conditions, resulting in a 2×2 factorial ANOVA.

If sample sizes are small or the parametric guidelines are not met, then a non-parametric test can be used. To compare the means of the two groups, Study 1 applied a Fischer's Exact Test, which is suitable for any sample size but typically used for smaller samples.

Finally, many modern studies choose to report effect size in addition to statistical significance. The effect size simply quantifies the difference in means between two

---

[12]Every statistical test has strict rules for its use—a detailed discussion of statistical methods is outside the scope of this chapter, so the experimenter should consult texts or other materials on inferential statistics prior to selecting a method for statistical analysis.

groups without complicating the analysis with sample sizes and statistical power concerns [87]. Effect size is calculated by simply dividing the difference in group means by the standard deviation.

Pilot and refine

If subjects are available, conduct one or more pilot studies prior to running actual trials. Pilot studies are extremely useful to researchers—they provide preliminary data on time, cost and practicality of the design. Pilots also offer the experimenter an indication of whether or not a hypothesized effect will be achieved. Finally, a pilot will allow the experimenter to improve upon the study design prior to conducting a full-scale research project. Studies 1 and 3 reported the procedures followed for pilot studies and how results were analyzed in order to improve upon the study design.

9.5.4 Cautions

Internal validity. One of the major threats to a between-subjects design is internal validity, or the potential for error arising from factors unrelated to the IV. In particular, individual difference factors such as age, gender, culture, social background, cognitive styles and tendencies, etc. Studies 1 and 3 were both between-groups designs and collected additional data to address internal validity concerns. As discussed earlier, this additional data can either be used as a pairing or blocking mechanism for assignment or may be used for additional correlation testing to aid in the interpretation of the participant responses. In either case, the goal is to identify any potential confounds in the analysis of performance.

Order or practice effects. Because participants in repeated-measures designs must complete multiple tasks, the primary danger is from order effects. As described earlier, order effects manifest either as improvement due to practice or worsening due to fatigue. Study 2 addressed practice effects by counterbalancing the order and assignment of vignettes across groups and subjects.

Realism. Although realism was stated earlier as a goal, it can also be a threat to the validity of the study. When basing a vignette on a real-life scenario, there is always a possibility that some of the subjects will recognize it and base their answers on their knowledge or recollection of events rather than on the evidence presented. Furthermore, even if unrelated to the actual event, it sometimes happens that real-world events interfere with the study by influencing participants. For example, a Study 3 vignette presented evidence surrounding an airline crash. If, during the study, such a crash would actually but unfortunately occur, the news media would undoubtedly discuss it and present theories as to its cause. Since it would be nearly impossible to sequester the subject pool from news or other information sources that may influence their thinking, experimenters are encouraged to run trials and collect data within as tight a time window as possible to minimize effect of outside influences.

9.6   Conclusion

The preceding sections, as well as others in this book, make a convincing case that cognitive heuristics often produce biased thinking that could lead even trained experts like intelligence analysts to biased judgments. That phenomenon, coupled with the disdain many expert analysts have for cumbersome, formal bias-reducing methodologies, is cause for concern among senior intelligence officials. The overarching goal of research is to investigate techniques to enable analysts to avoid the sorts of heuristics that lead to poor quality judgment without creating time-consuming process impediments. Such investigations should be conducted with experimental rigor and discipline so that the claims of efficacy and improved performance can be well-supported. This chapter presented a reference framework for experimentation based on several past studies, all of which contributed materially to these recommendations.

References

1. Baron J (2004) Normative models of judgment and decision making. In: Koehler DJ, Harvey N (eds) Blackwell handbook of judgment and decision making. Blackwell, London, pp 19–36
2. Commission on the Intelligence Capabilities of the United States Regarding Weapons of Mass Destruction (2004) Intelligence Reform and Terrorism Prevention Act of 2004
3. Cooper JR (2005) Curing analytic pathologies: pathways to improved intelligence analysis. Center for the Study of Intelligence, Washington, DC
4. Johnston R (2005) Analytic culture in the US Intelligence Community. Center for the Study of Intelligence, Washington, DC
5. Moore DT (2007) Critical thinking and intelligence analysis. National Defense Intelligence College, Washington, DC
6. National Commission on Terrorist Attacks upon the United States (2004) The 9/11 commission report. W.W. Norton & Company, New York, NY
7. U.S. Senate Select Committee on Intelligence (2004) Report on the U.S. Intelligence Community's Prewar Intelligence Assessments on Iraq
8. U.S. Office of Personnel Management (2009) Position classification standard flysheet for intelligence series, GS-0132. Government Printing Office, Washington, DC
9. Marrin S (2012) Evaluating the quality of intelligence analysis: by what (mis)measure? Intell Nat Secur 27(6):896–912
10. U.S. Joint Chiefs of Staff (2013) Joint publication 2.0: joint intelligence. Department of Defense, Washington, DC
11. U.S. Office of the Director of National Intelligence (2007) Intelligence Community Directive 203: Analytic Standards
12. Littlepage G, Robison W, Reddington K (1997) Effects of task experience and group experience on group performance, member ability, and recognition of expertise. Organ Behav Hum Decis Process 69(2):133–147
13. Fiedler FE (1994) Leadership experience and leadership performance. DTIC Document
14. Tversky A, Kahneman D (1974) Judgment under uncertainty: heuristics and biases. Science 185(4157):1124–1131
15. Elstein AS, Schwarz A (2002) Clinical problem solving and diagnostic decision making: selective review of the cognitive literature. BMJ 324:729–732
16. Pronin E, Lin DY, Ross L (2002) The bias blind spot: perceptions of bias in self versus others. Pers Soc Psychol Bull 28(3):369–381

17. Shanteau J (1992) The psychology of experts: an alternative view. In: Wright G, Bolger F (eds) Expertise and decision support. Plenum Press, New York, NY, pp 11–23
18. Stanovich KE, West RF (2008) On the relative independence of thinking biases and cognitive ability. J Pers Soc Psychol 94(4):672–695
19. Toplak ME, West RF, Stanovich KE (2011) The cognitive reflection test as a predictor of performance on heuristics-and-biases tasks. Mem Cognit 39(7):1275–1289
20. Heuer RJ (1999) Psychology of intelligence analysis. Center for the Study of Intelligence, Washington, DC
21. List of Cognitive Biases. Wikipedia n.d. Available from http://en.wikipedia.org/wiki/List_of_cognitive_biases
22. Lord CG, Ross L, Lepper MR (1979) Biased assimilation and attitude polarization: the effects of prior theories on subsequently considered evidence. J Pers Soc Psychol 37(11):2098–2109
23. Cook MB, Smallman HS (2008) Human factors of the confirmation bias in intelligence analysis: decision support from graphical evidence landscapes. Hum Factors 50(5):745–754
24. Ross L, Anderson CA (1982) Shortcomings in the attribution process: on the origins and maintenance of erroneous social assessments. In: Kahneman D, Slovic P, Tversky A (eds) Judgment under uncertainty: heuristics and biases. Cambridge University Press, Cambridge, MA, pp 129–152
25. Chapman LJ (1967) Illusory correlation in observational report. J Verbal Learn Verbal Behav 6(1):151–155
26. Klauer KC, Meiser T (2000) A source-monitoring analysis of illusory correlations. Pers Soc Psychol Bull 26(9):1074–1093
27. Garcia-Retamero R, Rieskamp J (2008) Adaptive mechanisms for treating missing information: a simulation study. Psychol Rec 58(4):547–568
28. Sanbonmatsu DM, Kardes FR, Herr PM (1992) The role of prior knowledge and missing information in multiattribute evaluation. Organ Behav Hum Decis Process 51(1):76–91
29. Kardes FR et al (2006) Debiasing omission neglect. J Bus Res 59(6):786–792
30. Troutman CM, Shanteau J (1977) Inferences based on nondiagnostic information. Organ Behav Hum Perform 19:43–55
31. LaBella C, Koehler DJ (2004) Dilution and confirmation of probability judgments based on nondiagnostic evidence. Mem Cognit 32(7):1076–1089
32. Windschitl PD, Wells GL (1998) The alternative-outcomes effect. J Pers Soc Psychol 75(6):1411–1423
33. Dougherty MR, Sprenger A (2006) The influence of improper sets of information on judgment: how irrelevant information can bias judged probability. J Exp Psychol Gen 135(2):262–281
34. Windschitl PD, Young M, Jenson M (2002) Likelihood judgment based on previously observed outcomes: the alternative-outcomes effect in a learning paradigm. Mem Cogn 30(3):469–477
35. Pallier G et al (2002) The role of individual differences in the accuracy of confidence judgments. J Gen Psychol 129(3):257–299
36. Björkman M, Juslin P, Winman A (1993) Realism of confidence in sensory discrimination: the underconfidence phenomenon. Percept Psychophys 54(1):75–81
37. Stanovich KE, West RF (1998) Individual differences in rational thought. J Exp Psychol Gen 127(2):161–188
38. Woodward B (2004) Plan of attack. Simon & Schuster, New York
39. Peterson JJ (2008) Appropriate factors to consider when expressing analytic confidence in intelligence analysis. Department of Intelligence Studies, Mercyhurst College
40. Kretz DR, Granderson CW (2013) An interdisciplinary approach to studying and improving terrorism analysis. In: 2013 IEEE international conference on intelligence and security informatics (ISI)
41. Lilienfeld SO, Ammirati R, Landfield K (2009) Giving debiasing away: can psychological research on correcting cognitive errors promote human welfare? Perspect Psychol Sci 4(4):390–398
42. Heuer RJ, Pherson RH (2010) Structured analytic techniques for intelligence analysis. CQ Press, Washington, DC

43. Heuer RJ (2005) How does ACH improve intelligence analysis? In: From the works of Richards J. Heuer, Jr
44. International Ergonomics Association. Definition and Domains of Ergonomics (2018) Available from https://www.iea.cc/whats/index.html
45. Fischhoff B (1975) Silly certainty of hindsight. Psychol Today 8(11):70
46. Halpern DF (1998) Teaching critical thinking for transfer across domains: dispositions, skills, structure training, and metacognitive monitoring. Am Psychol 53(4):449–455
47. Willingham DT (2007) Critical thinking: why is it so hard to teach? Am Educ 31(2):8–19
48. Larrick RP (2004) Debiasing. In: Koehler DJ, Harvey N (eds) Blackwell handbook of judgment and decision making. Blackwell Publishing, Oxford, UK, Malden, MA, pp 316–337
49. Arkes HR (1991) Costs and benefits of judgment errors: implications for debiasing. Psychol Bull 110(3):486–498
50. Lord CG, Taylor CA (2009) Biased assimilation: effects of assumptions and expectations on the interpretation of new evidence. Soc Pers Psychol Compass 3(5):827–841
51. Welsh MB, Begg SH, Bratvold RB (2007) Efficacy of bias awareness in debiasing oil and gas judgments. In: 29th Annual meeting of the cognitive science society, Cognitive Science Society, Nashville, TN
52. Intelligence Advanced Research Projects Activity. Sirius. n.d. Available from https://www.iarpa.gov/index.php/research-programs/sirius
53. Stapleton AJ (2004) Serious games: serious opportunities. In: Australian game developers conference, Academic Summit, Melbourne
54. Fischhoff B (1982) Debiasing. In: Kahneman D, Slovic P, Tversky A (eds) Judgment under uncertainty: heuristics and biases. Cambridge University Press, Cambridge, MA, pp 422–444
55. Lopes LL (1982) Procedural debiasing, OoN Research, Editor
56. Lopes LL (1987) Procedural debiasing. Acta Physiol (Oxf) 64(2):167–185
57. Nickerson RS (1998) Confirmation bias: a ubiquitous phenomenon in many guises. Rev Gen Psychol 2(2):175–220
58. Graber ML, Franklin N, Gordon R (2005) Diagnostic error in internal medicine. Arch Intern Med 165(13):1493–1499
59. Galinsky AD, Moskowitz GB (2000) Counterfactuals as behavioral primes: priming the simulation heuristic and consideration of alternatives. J Exp Soc Psychol 36(4):384–409
60. Mumma GH, Wilson SB (1995) Procedural debiasing of primacy/anchoring effects in clinical-like judgments. J Clin Psychol 51(6):841–853
61. Mussweiler T, Posten AC (2012) Relatively certain! Comparative thinking reduces uncertainty. Cognition 122(2):236–240
62. Kruger J, Wirtz D, Miller DT (2005) Counterfactual thinking and the first instinct fallacy. J Pers Soc Psychol 88(5):725–735
63. McKenzie CRM (1998) Taking into account the strength of an alternative hypothesis. J Exp Psychol Learn Mem Cogn 24(3):771–792
64. Folker RD (2000) Intelligence analysis in theater joint intelligence centers: an experiment in applying structured methods. Joint Military Intelligence College, Washington, DC
65. Sanna LJ, Schwarz N, Stocker SL (2002) When debiasing backfires: accessible content and accessibility experiences in debiasing hindsight. J Exp Psychol Learn Mem Cognit 28(3):497–502
66. Vallée-Tourangeau F, Beynon DM, James SA (2000) The role of alternative hypotheses in the integration of evidence that disconfirms an acquired belief. Eur J Cogn Psychol 12(1):107–129
67. Vallée-Tourangeau F, Villejoubert G (2010) Information relevance in pseudodiagnostic reasoning. In: 32nd Annual meeting of the cognitive science society, Portland, OR
68. Villejoubert G, Vallée-Tourangeau F (2012) Relevance-driven information search in "pseudodiagnostic" reasoning. Q J Exp Psychol 65(3):541–552
69. Treverton GF (2011) Foreward. In: Moore DT (ed) Sensemaking: a structure for an intelligence revolution. National Defense Intelligence College, Washington, DC
70. Hackman JR (2011) Collaborative intelligence: using teams to solve hard problems. Berrett-Koehler Publishers, San Francisco, CA

71. Markman KD, Hirt ER (2002) Social prediction and the "allegiance bias". Soc Cognit 20(1):58–86
72. Ely JW, Graber ML, Croskerry P (2011) Checklists to reduce diagnostic errors. Acad Med 86(3):307–313
73. Croskerry P (2003) Cognitive forcing strategies in clinical decisionmaking. Ann Emerg Med 41(1):110–120
74. Graber ML (2009) Educational strategies to reduce diagnostic error: can you teach this stuff? Adv Health Sci Educ 14:63–69
75. Haynes AB et al (2009) A surgical safety checklist to reduce morbidity and mortality in a global population. N Engl J Med 360(5):491–499
76. Trowbridge RL (2008) Twelve tips for teaching avoidance of diagnostic errors. Med Teach 30(5):496–500
77. Sherbino J et al (2011) The effectiveness of cognitive forcing strategies to decrease diagnostic error: an exploratory study. Teach Learn Med 23(1):78–84
78. Hernandez I, Preston JL (2013) Disfluency disrupts the confirmation bias. J Exp Soc Psychol 49(1):178–182
79. Kretz DR (2015) Strategies to reduce cognitive bias in intelligence analysis: can mild interventions improve analytic judgments?. The University of Texas at Dallas, Richardson, TX
80. Klayman J, Brown K (1993) Debias the environment instead of the judge: an alternative approach to reducing error in diagnostic (and other) judgment. Cognition 49(1–2):97–122
81. Smith JF, Kida T (1991) Heuristics and biases: expertise and task realism in auditing. Psychol Bull 109(3):472
82. Aronson E, Carlsmith JM, Ellsworth PC (1990) Methods of research in social psychology. McGraw-Hill, New York
83. Various (2007) Encyclopedia of social psychology. SAGE Publications, Inc., Thousand Oaks, Thousand Oaks, California
84. Rapoport A, Chammah AM, Orwant CJ (1965) Prisoner's dilemma: a study in conflict and cooperation, vol 165. University of Michigan press, Ann Arbor, IL
85. Smallman H (2008) JIGSAW–Joint intelligence graphical situation awareness web for collaborative intelligence analysis. In: Macrocognition in teams: theories and methodologies, pp 321–337
86. Ellis PD (2010) The essential guide to effect sizes: statistical power, meta-analysis, and the interpretation of research results. Cambridge University Press, Cambridge, MA
87. Coe R (2002) It's the effect size, stupid: what effect size is and why it is important

Chapter 10
Promoting Representational Fluency for Cognitive Bias Mitigation in Information Visualization

Paul Parsons

10.1 Introduction

Research throughout the past few decades has led to a considerable number of visualization techniques that can be used in any given context. For instance, when a designer wishes to visualize hierarchies, techniques such as treemaps, trees, or sunburst diagrams can be used; for networks, matrices and graphs can be used; for information flows, Sankey diagrams and decision trees can be used; for temporal changes, small multiples, streamgraphs, and spiral charts can be used; and so on. Research in the cognitive and learning sciences has consistently demonstrated that different representations (e.g. visualizations)[1] of the same data can influence cognition in significantly different ways [1, 31, 40]. While different representations can enhance cognitive performance by encouraging certain perceptual and cognitive operations, they can also elicit various biases in thinking and reasoning [22, 38, 40].

Representational biases manifest in two major ways: *constraints*—limits on what aspects of data can be expressed by a representation; and *salience*—how a representation facilitates processing of certain aspects of data, possibly at the expense of others [38]. Constraints arise due to the syntactical limitations of how graphical primitives are arranged in representational forms [31, 36], whereas salience arises from how easily information can be extracted from a representation. Such biases are not necessarily bad, as the value of constraints and salient features is context-dependent. For instance, when visualizing logic problems to support reasoning about sets, certain graphical constraints in Euler diagrams are beneficial, as intersecting circles can readily express underlying logical relationships [35]. When visualizing

---

[1] *Representation* and *visualization* are used interchangeably throughout when referring to external, visual representations of data. A discussion of internal representations is outside the scope of this paper.

P. Parsons (✉)
Purdue University, West Lafayette, IN, USA
e-mail: parsonsp@purdue.edu

© Springer Nature Switzerland AG 2018
G. Ellis (ed.), *Cognitive Biases in Visualizations*,
https://doi.org/10.1007/978-3-319-95831-6_10

networks to support reasoning about paths, matrices are limited in that they cannot directly express paths along multiple nodes, yet network diagrams do not have such a limitation [28]. However, matrices can make missing relations highly salient due to the existence of empty cells that can be perceived easily. Network diagrams, on the other hand, make such information only partially salient. Thus, the value of a representational bias (i.e., whether it is good or bad) depends on the context in which it is used. However, representational biases typically encourage thinking in certain ways at the expense of others, which can lead to the development of inaccurate or incomplete mental models. One way to mitigate this issue is to use multiple representations, thus providing different perspectives and encouraging multiple ways of thinking.

To work effectively with multiple representations, designers and users must be *fluent* in the various representations that are relevant for any given data and context. *Representational fluency* refers to knowledge and skills that involve being able to understand, use, create, evaluate, and translate between various representations. If individuals have *fluency* with multiple representational forms, they can employ appropriate practices that help mitigate the effects of representational biases. For example, when working with social network data, users can *translate* between a node-link diagram and an adjacency matrix depending on whether they want to identify paths in the network or the absence of relationships between two people. Representational fluency is considered necessary for professional discourse and practice in a number of fields including chemistry, physics, mathematics and biology. In this chapter, I argue that representational fluency should also be considered necessary for professional competence in information visualization and can be achieved through systematic training and education, in both formal and informal contexts. Thus, promoting representational fluency is a general strategy requiring concerted efforts of educators, researchers and practitioners.

Need for general strategies—Previous work in visualization has proposed general strategies for mitigating cognitive biases [7, 23, 30] as well as strategies for dealing with particular biases [8, 10]. While strategies focusing on specific visualizations, contexts, or biases are certainly useful and necessary, there is also a need for more general strategies. Extant scholarship on cognitive biases suggests that tackling specific biases, without complimentary general strategies, is not a sufficient approach, as biases often have multiple determinants. As Larrick [22] notes, "there is unlikely to be a one-to-one mapping of causes to bias, or of bias to cure". Thus, developing strategies for mitigating particular biases, while useful, does not constitute a sufficient research plan for dealing with cognitive biases in visualization. In this chapter, I propose that promoting representational fluency among visualization designers and users is one strategy that can help mitigate biases at a more general level. This strategy can complement techniques that are devised for dealing with specific biases, visualizations, or users.

10.2 Representational Fluency

Representational fluency has been studied in various STEM disciplines having a considerable interest in visualization— especially chemistry (e.g. [14, 20]), biology (e.g. [26, 39]), and physics (e.g. [15]). In these disciplines, many phenomena are not available for direct perception—e.g. molecules, atoms, proteins and forces. As a result, visual representations are essential for teaching, learning, communicating, and conducting research [13]. Interestingly, although visual representations are indispensable for working with abstract data, similar attention has not been paid to representational fluency and its attendant concepts in information visualization.

Studies show that experts are more fluent than novices with multiple representations in their disciplines [6, 18]. In fact, the degree to which individuals exhibit representational fluency is strongly correlated with their level of expertise. Although this has not been investigated in information visualization, presumably both expert visualization designers and users should have higher degrees of fluency than novices.

Extant scholarship on representational fluency does not point to universal agreement on the characteristics of fluency, nor does it reveal a coherent theoretical underpinning. Various scholars refer to fluency in different ways, sometimes treating it as synonymous with *representational competence*. While there is no well-established conceptual framework for discussing fluency, there is a strong consensus on some of its key features. For instance, most scholars appear to agree on the following requirements for fluency—being able to make sense of the meaning of representations; being able to translate between equivalent or complementary representations; being able to devise new representations that are contextually appropriate; being able to evaluate and critique existing representations; and understanding the functions of various representations and how and when they should be used [15, 25].

Hill et al. [15] recently reviewed the literature on representational fluency and suggest that contributions have been made from three related perspectives—(1) *representational competence*, (2) *meta-representational competence*, and (3) *meta-visualization*. Each of these perspectives is elaborated below. While there is considerable overlap among these perspectives, it is useful to understand their individual origins and contexts, to see how they may provide value for information visualization.

Representational Competence—Representational competence typically refers to the ability to comprehend and use a set of domain-specific representations. Representational competence comprises the ability to properly extract information from a representation—i.e., to understand its syntax and semantics. Individuals may have representational competence if they can 'see beyond' the surface-level characteristics of representations to their common underlying features, and are able to translate between different representations of the same data [21].

Meta-representational Competence—While representational competence refers to skills with a certain set of representations, *meta-representational competence* transcends this view, focusing on an approach where individuals understand the rationale for using particular representations and the design strategies used to create them [9, 15]. 'Meta' here is not used in a self-referential fashion; rather, it is used in the spirit of the original Greek meaning of "beyond" or "after"—e.g. as in metaphysics. Thus, meta-representational competence can be thought of as beyond simply competence with representations. Meta-representational competence is evidenced by skills such as *critiquing* visualizations to assess their suitability in particular contexts, *inventing* new visualizations and *describing* why and how a visualization works in a particular context.

Meta-visualization—Here, *visualization* refers to the process of making meaning from external representations. In this view, visualization is more of a cognitive phenomenon than an external artifact—visualization refers to not only an external representation, but to the internal representation (e.g. mental model) and the relationships between the two. This perspective has been promoted by Gilbert [12, 13] in science education and particularly in chemistry education. In this perspective, meta-visualization refers to "metacognition in respect of visualization" [12]. Gilbert argues that, just as there are generalized forms of spatial intelligence, memory, and thinking, there could similarly be generalized forms of meta-visualization. This perspective emphasizes the metacognitive processes and skills required to make meaning from external representations—e.g. the *monitoring* and *control* of what is being seen, what aspects should be retained, how they should be retained, and how they might be retrieved for later use. This perspective is different from the other two, as it very strongly focuses on the integration of external and internal representations, on cognitive processes such as mental modeling and mental simulation, and on the skills needed to have metacognitive proficiency in making meaning from external representations.

10.3 Implications for Visualization Research and Practice

The three perspectives described previously, reflect decades of work on representational fluency across various disciplines. These perspectives can provide a general framework from which to pursue and particularize representational fluency in information visualization. For instance, from the perspective of representational competence, representations for different types of data, users, domains and/or contexts could be compiled and characterized. To be representationally competent in one area requires an understanding of the syntax and semantics of the representations involved. To make meaning of a treemap visualization, for example, one must understand that shapes nested within each other communicate hierarchical levels; that the size of the shapes encodes a value; and, perhaps, that colors encode categorical features of the data. If these conventions are not understood, one cannot comprehend the treemap, and thus does not have competence with this particular representation. This could be extended to include a range of visualizations for hierarchical data. An individual should be able to look at an icicle plot, a sunburst diagram, a treemap and a node-link tree diagram and see beyond the surface level marks and encodings, being able to recognize common features in the underlying data. They would be able to identify the same kinship relations in the different representations—e.g.

parent-child, sibling, ancestor and descendant relations. They would know that some representations encode parent-child relationships explicitly with lines, while others encode them implicitly using features such as position, overlap or containment. Furthermore, given a treemap, they would be able to decode it and express the same data using an icicle plot.

An individual with meta-representational competence should be able to critique a visualization, describing why it is or is not appropriate in a given context, and should be able to devise a new representation based on the data and users' tasks. While representational competence refers to the *what* and *how* of representations— e.g. what do they represent and how is it done, meta-representational competence refers to the *why* of a representation—e.g. why it works the way it does and why it is appropriate or inappropriate for the data and context. Individuals who are meta-representationally competent should be comfortable answering these types of 'why' questions in addition to 'how' and 'what' questions—e.g. why is a heatmap or parallel coordinates plot appropriate in a given context, how can one be constructed from the other, and so on.

The meta-visualization perspective is perhaps the least straightforward of the three perspectives. This perspective requires a more holistic lens, examining the distributed cognitive system comprising both internal and external representations and processes. Furthermore, it requires examining the metacognitive skills that operate on those processes. From this perspective, individuals should be able to articulate what kind of knowledge they are acquiring while viewing and interacting with visualizations, how and why they are storing various aspects and views on the data in memory, how they are relating this new knowledge to existing knowledge, and how they might retrieve it for later use for problem solving or other activities. Although Gilbert [12] suggests that meta-visualization can be assessed through various verbal protocols (e.g. think-aloud) and interviews, no detailed assessment methods have been devised. Further research is needed to determine how meta-visualization could be assessed in the context of information visualization.

10.3.1 Developing Representational Fluency

The strategy being proposed here will not be very effective if implemented only in specific cases to deal with specific biases. Although individual designers and users can indeed develop representational fluency, which should help mitigate potential biases that may arise, the ideal solution is for representational fluency to be promoted systematically during visualization education, training and practice. This suggestion is not unattainable, as it as already an accepted expectation in other disciplines such as chemistry, physics, biology and mathematics. For instance, for professional chemists, representational fluency is an inseparable aspect of their expertise.

An important caveat here is that we cannot always expect users of visualizations to be experts. As information visualization becomes more prevalent in everyday contexts, more non-experts are exposed to visualization techniques on a regular basis. For instance, as data journalism grows in popularity, more online news sources integrate visualizations into their news stories, which are read by the general public. While theoretically possible to train the general public to develop representational fluency with common visualization techniques (after all, most students are taught how to read and use bar and line charts, scatterplots and other common techniques in school), it is not reasonably practicable in the near future.

A more reasonable expectation is that visualization designers develop a high degree of representational fluency during their training. As a result, designers could anticipate when various representational biases may manifest themselves, and integrate deliberate strategies into their visualizations to help mitigate the biases. For instance, consider a designer wanting to visualize temporal change. If they know that an animated chart may have a representational bias, in that it is limited to expressing data only at particular points in time, they may choose to use a small multiples technique instead, which does not share the same representational bias [3]. Alternatively, the designer may implement an option for users to interactively translate between the animated chart and the small multiples view (which also has cognitive benefits other than mitigating biases; see [32]). Because of the designer's representational fluency, they implement this option deliberately, knowing that it can help mitigate biases. Furthermore, depending on the context, the visualization tool may even encourage users to translate between the representations at certain points in time. With ongoing advances in intelligent mixed-initiative systems, such a prospect may not be so unlikely in the near future.

It is worth noting here that in order to most effectively mitigate cognitive biases, representational fluency must complement established knowledge on perception, cognition, decision making, semiotics, interaction design, visual encodings and other relevant topics. Representational fluency is not a panacea for all problems related to cognitive biases in information visualization.

10.3.2 Effect on Cognitive Processing

Much of the theoretical basis of cognitive debiasing suggests that successful strategies encourage individuals to move from surface-level to deeper-level thinking [22]. This can be viewed as a shift from 'System 1' to 'System 2' thinking in the language of Kahneman [19], or from 'experiential' to 'reflective' modes of cognition in the language of Norman [27]. Whatever the language, the intention is to shift cognitive processing from the fast, intuitive, unconscious mode to the slow, reflective, conscious mode. This is somewhat at odds with typical goals espoused in the visualization literature—namely, to offload as much cognitive processing as possible onto the perceptual system and onto external artifacts (e.g. visualization tools and computational processing).

Although it is generally desirable to offload cognitive processing when working with visualizations, mitigating cognitive biases may be an area in which it is beneficial to place more burden of cognitive processing onto users. However, increasing cognitive burden must be done in a principled fashion, as not all cognitive burden is beneficial. For instance, trying to make sense of a network visualization that is extremely complex, with considerable occlusion of nodes and edges, will certainly increase cognitive burden—yet this increase is not beneficial and could be avoided with better design. However, after working with one visualization for a while, translating to an alternative visualization may lead to increased cognitive burden—yet, this increase can be beneficial, as it forces the user into a more reflective mode of cognitive processing in which critical questions may be asked of the underlying data. Another strategy is to design interactions to deliberately influence cognitive processing, increasing the cognitive burden where designers deem appropriate (see [29, 34]). Indeed, strategies for manipulating cognitive effort through interactive interface design have been studied in the context of educational and learning technologies for many decades now (e.g. [5, 17, 33]).

Evidence for the benefits of deeper cognitive processing in cognitive bias mitigation can be found in the literature on cognitive debiasing. For instance, research has shown that counter-explanation, having individuals devise alternative explanations to observations, can help mitigate known biases, such as the explanation bias [2] and the hindsight bias [4]. Studies suggest that counter-explanation tasks may be beneficial by disrupting individuals' focal hypotheses and engendering more thorough and careful thinking about the phenomena under investigation [16]. Although representational fluency is not the same as devising alternative explanations, seeing multiple representations of the same data may effect the same cognitive processes responsible for disrupting focal hypotheses. Other known strategies for mitigating biases, such as reference class forecasting [11], also rely on engaging individuals in deeper cognitive processing to be successful. As the strategy of deliberately engaging users in deeper cognitive processing has not traditionally been an area of focus for the information visualization community, there is a need for a research agenda that outlines the main challenges to be overcome.

10.3.3 Preliminary Research Agenda

Based on the work above, five broad challenges are enumerated for a research agenda focusing on representational fluency. These five challenges are not intended to be entirely orthogonal or exhaustive. It is worth noting that these challenges are very general and could likely be broken down into more specific sub-challenges. However, at this point, they give structure to a wide range of challenges in this area and can help direct future research. Future work will likely identify more specific challenges and appropriate methodologies for dealing with them. Based on work in other disciplines concerned with representational fluency and interactive visualizations, along with existing research on cognitive bias mitigation, these five points set the stage for a more elaborate research agenda to unfold in the future.

1. **Identify a core set of representations in which all visualization professionals should be competent**. This is a difficult challenge, as there are currently many dozens of existing visualization techniques and new ones are continually being devised. Additionally, not all visualizations are appropriate in all contexts, and some visualizations are intended for very particular contexts. It may not be possible to identify a universally agreed-upon set of representations. However, without at least a rough set of common representations, it is difficult to promote and assess fluency in them. It may be the case that core sets of representations are identified for different contexts, users and data, and fluency in one or more sets can be promoted and assessed.

2. **Identify pedagogical practices that promote representational fluency**. Without concerted efforts on the part of visualization educators, it is unlikely that designers can develop fluency with various representations. Educators need to develop pedagogical strategies and practices for promoting representational competence, meta-representational competence, and meta-visualization. Although work has been done in other disciplines, it is not necessarily transferable to information visualization. Well-trained visualization designers should be able to understand, for example, the semantics of various encodings in different representations, their particular representational biases, how and why they were created and when they are most appropriate to be used. They should also understand which visualizations can complement each other, and when and how users should be able to translate between them.

3. **Develop ways of assessing representational fluency**. Without both formal and informal ways of assessing individuals' representational fluency, pedagogical practices go only so far. There is a need for the development of formally administered methods of testing representational fluency, as well as means of self-assessing fluency. For example, surveys such as the one by Hill et al. [15] could be developed for common visualization techniques. Other strategies, such as protocol analysis and eye-tracking [37], could also be explored. Educators could devise standardized tests in which various aspects of representational fluency can be assessed. To emphasize the more designerly aspects of visualization practice, various design challenges could be given. Classroom practices that encourage critical reflection, such as design critiques, could be employed both formally and informally to assess the development of students' representational fluency.

4. **Investigate strategies for appropriately engendering deeper cognitive processing**. As discussed above, research on cognitive debiasing consistently shows that effective interventions tend to shift individuals' thinking from a surface, unconscious level to a deeper, conscious level. Various strategies for implementing this in a visualization context can be explored. For instance, the representations that are made available to users, and the sequences in which they are made available could be manipulated; various interactions could be made available or unavailable to users at different points in time to encourage different cognitive operations; even micro-level aspects of interactions can be manipulated to promote more reflective thinking (e.g. see [24]). To tackle this challenge,

the visualization community could advantageously borrow strategies from the instructional design and learning technologies literature.

5. **Test effects on cognitive biases in various experimental settings**. Although promoting representational fluency is a general strategy, which should have effects across a range of biases, it is still important to test bias mitigation with specific biases and visualizations. Experiments could be devised where individuals that are known to have representational fluency in at least some subset of representations (as determined by assessments mentioned in challenge 3 above) are given visualizations with known representational biases, and are given the means to interactively translate between representations while performing tasks. Various strategies devised in response to challenge 4 above could also be tested, shedding light on both the strategies of designers and the effects on users.

10.4 Summary

The development of representational fluency by visualization designers and users is one strategy for mitigating cognitive biases when working with visualizations. As representational fluency is a well-established expectation for professionals in a number of disciplines, it is not unreasonable to have the same expectation in information visualization. Furthermore, representational fluency is a serious topic for research and scholarship in other disciplines, and should be too in information visualization. Establishing representational fluency among visualization professionals will require a concerted effort on the part of educators, researchers and practitioners, and will likely have multiple benefits beyond mitigating cognitive biases. For instance, representational fluency can lead to better communication among researchers and practitioners; better trained designers who know when and how to implement particular visualizations and interactions; and users who are more visualization literate, which can be of benefit across a wide range of data-driven activities.

References

1. Ainsworth S (2006) DeFT: a conceptual framework for considering learning with multiple representations. Learning and Instruction 16(3):183–198
2. Anderson CA, Sechler ES (1986) Effects of explanation and counterexplanation on the development and use of social theories. J. Personality Social Psych 50(1):24
3. Archambault D, Purchase H, Pinaud B (2011) Animation, small multiples, and the effect of mental map preservation in dynamic graphs. IEEE Trans Visualization Comput Graphics 17(4):539–552
4. Arkes HR, Faust D, Guilmette TJ, Hart K (1988) Eliminating the hindsight bias. J Appl Psych 73(2):305
5. Baker M, Lund K (1997) Promoting reflective interactions in a CSCL environment. J Comput Assisted Learning 13(3):175–193

6. Chi MTH, Feltovich PJ, Glaser R (1981) Categorization and representation of physics problems by experts and novices. Cognitive Sci 5(2):121–152
7. Correll M, Gleicher M (2014) Bad for Data, Good for the Brain: knowledge-first axioms for visualization design. In: Proceedings of the 1st workshop on dealing with cognitive biases in visualisations (DECISIVe 2014)
8. Dimara E, Dragicevic P, Bezerianos A (2014) Accounting for availability biases in information visualization. In: Proceedings of the 1st workshop on dealing with cognitive biases in visualisations (DECISIVe 2014)
9. DiSessa AA (2004) Metarepresentation: native competence and targets for instruction. Cognition Instruction 22(3):293–331
10. Dragicevic P, Jansen Y (2014) Visualization-mediated alleviation of the planning fallacy. In: Proceedings of the 1st workshop on dealing with cognitive biases in visualisations (DECISIVe 2014)
11. Flyvbjerg B (2008) Curbing optimism bias and strategic misrepresentation in planning: reference class forecasting in practice. Eur Plann Stud 16(1):3–21
12. Gilbert JK (2005) Visualization: a metacognitive skill in science and science education. In: Visualization in Science Education, Springer, pp 9–27
13. Gilbert JK (2008) Visualization: an emergent field of practice and enquiry in science education. Theory and Practice in Science Education, Visualization, pp 3–24
14. Grove NP, Cooper MM, Rush KM (2012) Decorating with arrows: toward the development of representational competence in organic chemistry. J Chem Educ 89(7):844–849
15. Hill M, Sharma M, O'Byrne J, Airey J (2014) Developing and evaluating a survey for representational fluency in science. Int J Innovation Sci Mathe Educ 22(6):22–42
16. Hirt ER, Markman KD (1995) Multiple explanation: a consider-an-alternative strategy for debiasing judgments. J Personality Soc Psych 69(6):1069–1086
17. Jackson SL, Krajcik JS, Soloway E (1998) The design of guided learner-adaptable scaffolding in interactive learning environments. In: Proceedings of the ACM conference on human factors in computing systems (CHI '98) pp 187–194
18. de Jong T, Ferguson-Hessler MGM (1991) Knowledge of problem situations in physics: a comparison of good and poor novice problem solvers. Learning Instruction 1(4):289–302
19. Kahneman D (2013) Thinking, fast and slow. Penguin Books Ltd., London
20. Kozma R, Russell J (2005) Students becoming chemists: developing representational competence. Visualization in Science Education pp 121–145
21. Kozma RB, Russell J (1997) Multimedia and understanding: expert and novice responses to different representations of Chemical phenomena. J Res Sci Teaching 34(9):949–968
22. Larrick RP (2004) Debiasing. In: Blackwell Handbook of Judgment and Decision Making, Blackwell Publishing, pp 316–337
23. Leuven KU, Verbeiren T, Leuven KU, Aerts J (2014) A pragmatic approach to biases in visual data analysis. In: Proceedings of the 1st workshop on dealing with cognitive biases in visualisations (DECISIVe 2014)
24. Liang HN, Parsons P, Wu HC, Sedig K (2010) An exploratory study of interactivity in visualization tools: 'Flow' of interaction. J. Interactive Learning Res 21(1):5–45
25. Nathan MJ, Alibali MW, Masarik K, Stephens AC, Koedinger KR (2010) Enhancing middle school students representational fluency: a classroom-based study. (WCER Working Paper No 2010-9) Retrieved from University of WisconsinMadison, Wisconsin Center for Education Research website: http://www.wcerwiscedu/publications/workingpapers/papersphp
26. Nitz S, Tippett CD (2012) Measuring representational competence in science. In: EARLI SIG 2 Comprehension of Text and Graphics, pp 163–165
27. Norman DA (1993) Things that make us smart: defending human attributes in the age of the machine. Addison-Wesley
28. Novick L, Hurley SM (2001) To matrix, network, or hierarchy: that is the question. Cognitive psychology 42(2):158–216
29. Parsons P, Sedig K (2014) Distribution of information processing while performing complex cognitive activities with visualization tools. In: Huang W (ed) Handbook of human-centric visualization, Springer, New York, chap 28, pp 693–715

30. Pohl M, Winter LC, Pallaris C, Attfield S, Wong BLW, (2014) Sensemaking and cognitive bias mitigation in visual analytics. Proceedings -, (2014) IEEE joint intelligence and security informatics conference. JISIC 2014:323. https://doi.org/10.1109/JISIC.2014.68
31. Scaife M, Rogers Y (1996) External cognition: how do graphical representations work? Int J Human-Comput Studies 45(2):185–213
32. Sedig K, Parsons P (2013) Interaction design for complex cognitive activities with visual representations: a pattern-based approach. AIS Trans on Human-Comput Interact 5(2):84–133
33. Sedig K, Klawe M, Westrom M (2001) Role of interface manipulation style and scaffolding on cognition and concept learning in learnware. ACM Trans Comput-Human Interact (TOCHI) 8(1):34–59
34. Sedig K, Parsons P, Dittmer M, Haworth R (2014) Human-centered interactivity of visualization tools: micro- and macro-level considerations. In: Huang W (ed) Handbook of human-centric visualization, Springer, New York, chap 29, pp 717–743
35. Stenning K, Lemon O (2001) Aligning logical and psychological perspectives on diagrammatic reasoning. Artif Intell Rev 15(1–2):29–62
36. Stenning K, Oberlander J (1995) A cognitive theory of graphical and linguistic reasoning: logic and implementation. Cognitive Sci 19(1):97–140
37. Stieff M (2011) Improving representational competence using molecular simulations embedded in inquiry activities. J Res Sci Teaching 48(10):1137–1158
38. Suthers DD (1999) Representational bias as guidance for learning interactions: a research agenda. New computational technologies to support learning, exploration and collaboration, Artificial Intelligence in Education Open learning environments, pp 121–128
39. Wilder A, Brinkerhoff J (2007) Supporting representational competence in high school biology with computer-based biomolecular visualizations. J Comput Mathe Sci Teaching 26:5–26
40. Zhang J (1997) The nature of external representations in problem solving. Cognitive sci: a Multidisciplinary J 21(2):179–217

Chapter 11
Designing Breadth-Oriented Data Exploration for Mitigating Cognitive Biases

Po-Ming Law and Rahul C. Basole

11.1 Introduction

Exploratory data analysis empowers users to discover the unanticipated from data. In search of insights, users examine a body of information and articulate questions about the data in an iterative fashion [9]. Aside from being loaded with a vast amount of new information, users have to make a series of complex decisions while navigating through data: what questions should I ask? which piece of information should I examine to answer my questions? As users often do not have good knowledge about the data they are exploring, making these decisions is difficult. Unconscious shortcuts are often applied in making these decisions, letting heuristics drive users' exploration. While heuristics maintains analysis flow by shielding users from making a conscious effort in every step of data exploration, a biased exploration path might hinder insight generation and lead to confirming hypotheses erroneously.

A characteristic in a biased exploration path is lack of breadth. Users may be fixated on a question in the early stage of exploration and, as a result, the coverage of a dataset is constrained. For instance, psychology studies showed that people tend to search for information which confirms pre-existing hypotheses (confirmation bias) [5]. People also tend to associate higher importance to things they can recall better and potentially explore the related information more (availability heuristics) [10]. In addition, data analysis is often initiated by formulating a goal or an anchor and, once the anchor is set, people tend to not deviate too far from it (anchoring bias) [11].

We argue that these bias and heuristics can be alleviated by breadth-oriented data exploration. Take for example anchoring bias, when a user is looking for a car to purchase, they might start by searching for cars with a low price. Without considering other car attributes, they tends to pay excessive attention to the cheap vehicles and

P.-M. Law (✉) · R. C. Basole
Georgia Institute of Technology, 85 5th St NW, Atlanta, GA, USA
e-mail: pmlaw@gatech.edu

R. C. Basole
e-mail: basole@gatech.edu

© Springer Nature Switzerland AG 2018
G. Ellis (ed.), *Cognitive Biases in Visualizations*,
https://doi.org/10.1007/978-3-319-95831-6_11

makes a sub-optimal purchase decision. A breadth-oriented exploration system like Voyager [14] can expose users to the other car attributes, assist users with assessing alternatives which are not as cheap but have other desirable properties and hence help users adjust from the bias.

While systems focusing on breadth-oriented exploration start to emerge, how to design such systems is yet to be explored. Voyager [14] provides insights into designing breadth-oriented systems for exploring tabular data. Yet, a large variety of data types are involved in exploratory data analysis in different domains. Developing new breadth-oriented systems for different data types would be challenging if the design process is not informed by any guidelines.

In this chapter, we contribute three considerations involved in designing systems which support breadth-oriented data exploration. To demonstrate the utility of these design considerations, we illustrate a hypothetical system which facilitates breadth-oriented exploration of dynamic networks. Finally, we discuss the challenge, the opportunities and the future work in advancing the science of breadth-oriented exploration.

11.2 The Information Space Model of Breadth-Oriented Exploration

To elucidate the process of breadth-oriented exploration and facilitate the discussion on our proposed design considerations, we present the information space model of breadth-oriented exploration.

In the information space model, each dataset has its own information space (Fig. 11.1a). An information space is a set of information pieces which can be derived from the data. For example, in the well-known car dataset [2], an information piece can be "the dataset covers cars produced between 1970 and 1982" or "acceleration seems to be normally distributed". Some of these information pieces are deemed as insights by a user while some of them are not. Insights are the information pieces which present meaningful knowledge to users, help make decisions (e.g. which car to buy) and validate hypotheses. As insights are user-defined, which information pieces correspond to insights vary among users.

During exploratory data analysis, users expand their coverage of the information space as they gather more information pieces (Fig. 11.1b). Due to the heuristics they unconsciously resort to during opportunistic exploration, the covered set of information pieces might be biased (Fig. 11.1c). The consequence is that we might not be able to reach some of the information pieces which generate insights. For example, users may be fixated on the price of cars and do not pay attention to the options that are not as cheap but have other good qualities. This can also happen when users constrain themselves to a small set of information pieces in an attempt to confirm their hypotheses and do not explore the alternative hypotheses. Through *active system feedback*, a system that supports breadth-oriented exploration grants users access to the information pieces that they would have missed if they had explored the

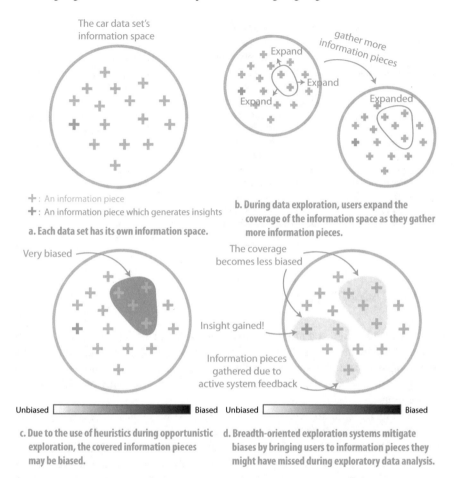

+ : An information piece
+ : An information piece which generates insights

a. Each data set has its own information space.

b. During data exploration, users expand the coverage of the information space as they gather more information pieces.

Unbiased [gradient bar] Biased   Unbiased [gradient bar] Biased

c. Due to the use of heuristics during opportunistic exploration, the covered information pieces may be biased.

d. Breadth-oriented exploration systems mitigate biases by bringing users to information pieces they might have missed during exploratory data analysis.

Fig. 11.1 Four characteristics of the information space model

information space alone, thus helping users glean insights they would have missed. This may lead to mitigating biases of their exploration (Fig. 11.1d).

11.3 Three Considerations for Designing Breadth-Oriented Data Exploration

In this section, we offer three considerations involved in designing systems that support breadth-oriented exploration. These design considerations are unit of exploration, user-driven versus system-driven exploration and related versus systematic exploration (Table 11.1).

Table 11.1 The three considerations for designing breadth-oriented data exploration

Unit of exploration	User-driven versus system-driven exploration	Related versus systematic exploration
Types: dimension or data case **Uses**: (1) facilitating the design of breadth-oriented exploration systems (e.g. when the unit of exploration is dimension, systems can provide dimension coverage information to create awareness of biases). (2) quantifying the effectiveness of a system in promoting breadth-oriented exploration	**User-driven**: a breadth-oriented system creates awareness of a biased exploration path and users expand the information space in a less biased manner as they become aware of their biases **System-driven**: the system actively presents information extracted from the dataset to expand the information space in a less biased manner	**For system-driven exploration**: **Related**: as users express their interests, the system presents something related but different **Systematic**: to complete an analysis, users need to finish some steps in a predefined exploration path

11.3.1   Unit of Exploration

A unit of exploration is a tool for thinking about the breadth of users' exploration. It can be a dimension or a data case. The use of this tool is twofold - it facilitates designing a system that supports exploration in greater breadth and it serves as a metric for quantifying the effectiveness of a system in promoting breadth-oriented exploration.

If the unit of exploration is a dimension, designers may consider providing dimension coverage information to help users keep track of how much of the information space they have explored thus far. A dataset can have hundreds of dimensions. While analyzing a dataset, users can explore any combinations of these dimensions. Hence, the space of dimension combination is huge. If users do not know what combinations have or have not been seen, they may fixate on certain combinations due to biases, limiting the breadth of their exploration. Showing dimension coverage information creates an awareness of what has not been explored to steer users towards the unexplored dimensions and combinations. This is supported by Sarvghad et al. [8] who demonstrated that incorporating dimension coverage information into an interface increases the breadth of exploration without sacrificing depth. Besides showing information about users' provenance of exploration, a breadth-oriented system can actively present more dimensions to users while they are exploring the data. For instance, as users indicate their interests in one dimension, Voyager [14] expands the covered information pieces in the information space by displaying statistical charts with unseen dimensions. Theoretically, similar principles can be applied when a data case is a unit of exploration. For example, a breadth-oriented system can be designed to help users track what data cases (e.g. documents) have already been explored and what have not. It is similar to email applications which allow us to mark an email as "read" and hence making us more aware of the unread emails.

Apart from helping designers think about how an interface should be designed to encourage breadth-oriented exploration, unit of exploration also provides a simple metric of breadth of exploration. When designers evaluate the effectiveness of their systems in encouraging broad exploration, they can measure how many dimensions or data cases the subjects covered while using their systems in comparison with other tools which do not support breadth-oriented exploration. The assumption is that the more the units of exploration covered, the greater the breadth of exploration. While this metric is simple, it might not be as reliable as other metrics such as the number of findings which indicate a new line of inquiry during exploratory analysis [8].

11.3.2 User-Driven Versus System-Driven Exploration

Another question in designing a breadth-oriented exploration system is whether the expansion of information space is user-driven or system-driven.

User-driven systems create an awareness that users' exploration might be biased. Knowing that their exploration has been biased, users can expand the information space in a less biased way. Albeit focusing on cohort selection rather than data exploration, adaptive contextualization [3], for example, illustrates how creating awareness of biases can help users adjust from the biases. As users select a patient cohort, their system presents information about the distribution of the selected cohort. Being aware of the biases in the distribution, users tend to adjust the criteria for cohort selection. In the context of data exploration, system designers can create an awareness of a biased exploration path by presenting information about what and how much has been explored so far and even what other people have explored (like scented widget [13]).

In contrast, system-driven exploration actively expand the information space while users are exploring the data. As users express their interests in something (e.g., a dimension), these systems actively present something related but different. With Voyager [14], users express their interest in some dimensions and the system displays charts with the selected dimensions as well as an unseen dimension. This technique is widely adopted by the graph visualization community (e.g. [1, 4, 12]). For example, with Apolo [1], users start the exploration by putting some nodes into groups. The system then searches for some nodes which are related to the group from a network with thousands of nodes and present them to users.

The key difference between user-driven exploration and system-driven exploration lies in what information is presented to users. In user-driven exploration, a breadth-oriented system presents information about users' exploration history to create an awareness that users' exploration may be biased. Users rather than the system are responsible for adjusting their exploration. In system-driven exploration, the system presents extra information extracted from the data to directly expand the coverage of the information space. Adjusting users' exploration is the responsibility of the system rather than users.

11.3.3   Related Versus Systematic Exploration

If expansion of the information space is driven by systems, two more considerations are involved: whether the expansion is based on users' interests (related exploration) or the expansion is systematic and not related to users' interests (systematic exploration).

Both Voyager [14] and Apolo [1], mentioned previously, falls into the category of related exploration. There are two sides of this same coin: presenting something *similar* to but *different* from users' interests. By showing something *similar* (like Voyager [14] which shows statistical charts related to the variables users are interested in), these systems maintain users' theme and flow of analysis. By showing something *different* from what is indicated by users (like Voyager [14] which shows some unseen dimensions in the recommended statistical charts), these systems expand the information space, bringing users to information pieces which they would otherwise have missed. This approach bears some resemblance to Amazon, which based on users' purchase history, recommends products that users might have missed.

For systematic exploration, systems expand the information space in a systematic way. A predefined path of exploration is planned out prior to analysis. In order to complete an analysis, users need to finish all the steps in the predefined path. For example, Perer and Shneiderman [7] concluded that there are 7 steps in social network analysis from their experience with domain experts. While this approach lacks flexibility of freely exploring the data, it ensures that users will not miss any important information pieces. As users' exploration is driven by predefined paths rather than heuristics, it is likely an effective approach to mitigating biases in data exploration. Yet, designing the predefined exploration path is clearly not an easy task.

11.4   Application of the Three Design Considerations

To demonstrate the utility of the three design considerations, we propose a hypothetical system that facilitates breadth-oriented exploration of dynamic social networks. In conducting the design study, we first consider a common task involved in social network analysis (SNA). We then illustrate a motivating usage scenario of the proposed system. Finally, we explain how the system is designed based on our three considerations.

11.4.1   Task Analysis

One common task in social network analysis is to understand the temporal characteristics of different groups of ego-networks. In an online social network, each person in this network is connected to many other people (e.g. their friends). An ego-network consists of a focal node (a person) and the nodes which are directly connected to it (e.g. the person's friends). These ego-networks are dynamic in nature because the set of nodes that are connected to a focal node may change over time.

Consider Mary, a healthcare researcher who wants to explore a dynamic social network. In the healthcare domain, many researchers are interested in knowing whether a larger social network leads to better health [6]. In Mary's dataset (Fig. 11.2), each node is described by a label (either healthy or unhealthy) and a feature vector that quantifies the temporal characteristics of the node's ego-network. Suppose there are three elements in this feature vector: (i) average size of the node's ego-network (average number of friends connected to the focal node), (ii) a metric that indicates how fluctuating the ego-network size is and (iii) a metric that indicates the average number of clusters in the ego-network. Our goal is to design a system that enables Mary to make sense of the temporal characteristics of the ego-networks of healthy and unhealthy people. This system should encourage Mary to explore her dataset in breadth and mitigate her biases during data exploration.

11.4.2 Usage Scenario

Mary recalls that many of her friends who look healthy have a large social circle. She has an intuitive feeling that a larger social circle leads to better health (availability heuristics). She initiated her analysis with the system by searching for evidence to confirm her hypothesis (confirmation bias). To do so, she asks the system what are

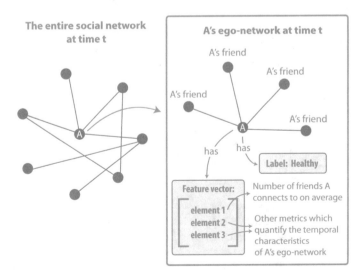

Fig. 11.2 Mary's dynamic network dataset. Note that both the entire network (left) and A's ego-network (right) change over time

the distinguishing features of the healthy group compared with the unhealthy group (Fig. 11.3a). Using data mining techniques, the system tells her that the healthy group, in general, has a larger social network, while the unhealthy group, in general, has a smaller social network (Fig. 11.3b). A system that does not support breadth-oriented exploration will stop the analysis here, causing Mary to believe that her hypothesis is true.

Knowing that Mary is interested in ego-networks with a large average network size, the proposed breath-oriented exploration system ranks the ego-networks based on their average network size (i.e. the average number of nodes to which a focal

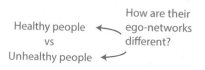

a. Mary asks the system how are healthy people's ego-networks different from unhealthy people's ego-networks?

b. The system responds by telling Mary that in general healthy people have a larger ego-network/ social circle than unhealthy people.

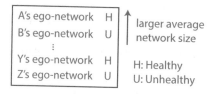

c. Knowing that Mary is interested in network size, the system ranks the ego-networks based on their average network size.

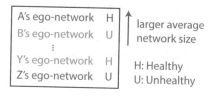

d. Some people who have a large network are unhealthy while some people who have a small network are healthy.

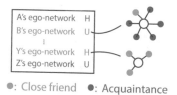

•: Close friend   ●: Acquaintance

e. Mary observes that healthy people who have a small network connects to more close friends than acquaintances while unhealthy people who have a large network connects to more acquaintances than close friends.

More close friends

↓ leads to

More social support

f. Mary suspects that connection with close friends also play a role in determining health. As it is easier to get support from close friends, she hypothesizes that social support is a more important factor in determining health.

Fig. 11.3 A usage scenario of the proposed breadth-oriented exploration system

node is connected) (Fig. 11.3c). Mary notices that some people who have a large average network size are unhealthy and some people who have a small average network size are healthy (Fig. 11.3d). She visualizes these ego-networks using a node-link diagram. She observes that the unhealthy people who have a large network, in general, connect to more acquaintances than close friends and the healthy people who have a small network, in general, connect to more close friends than acquaintances (Fig. 11.3e). As it is easier to get social support from close friends than acquaintances, Mary has an alternative hypothesis: social support rather than social network size is a more important factor in determining health (Fig. 11.3f). The breadth-oriented system successfully helps Mary move beyond the original line of inquiry to consider an alternative hypothesis.

11.4.3   Designing Based on the Three Considerations

The three design considerations are applied in designing the hypothetical system.

Unit of exploration: In designing the system for exploring Mary's dynamic network, we can choose node, link or dynamic ego-network as a unit of exploration. To capture the temporal characteristics better, we choose *dynamic ego-network* as a unit. When Mary expresses her interests in the ego-networks with a large average size, the system presents a ranked list of dynamic ego-networks based on their average network size.

System-driven versus user-driven exploration: Rather than creating an awareness of a biased exploration path and letting Mary refine her exploration on her own, the proposed system achieves breadth-oriented exploration by actively presenting an ordered list of ego-networks that Mary may be interested in. As the system actively expands the information space by showing information pieces that users might have missed, the system is designed to be *system-driven*.

Related versus systematic exploration: The system creates a ranked list of related ego-networks when Mary demonstrates her interests in ego-networks with a large network size. *Related exploration* is adopted when designing the system.

11.5   Discussion

We end this chapter by discussing the challenge, opportunities and future work in advancing the science of breadth-oriented exploration.

Information overload is a clear *challenge*: An obvious concern with breadth-oriented exploration is information overload. The extra information presented by breadth-oriented systems in users' course of exploration requires extra cognitive effort to process. Worse still, these systems may present irrelevant information to users. The question we need to answer is how to ensure the relevance of the information to be presented, especially in system-driven exploration in which a system actively present information pieces from the dataset. A potential solution is to create models of users based on their exploration history. The model may contain information about a user's interests and what they have explored so far. This approach is similar to recommender systems that infer what users are interested in based on their profiles, search history or purchase history. With users' models, breadth-oriented systems can predict what information users are interested in, prune the search space of information to be presented and provide more relevant information.

Breadth-oriented exploration offers *opportunities* **in the big-data era**: Beyond cognitive bias, breadth-oriented exploration presents excellent opportunities in the era of big data. Consider a data table with millions of rows. It is likely that users will miss a lot of valuable insights during exploratory data analysis. Due to information overload and the complex decisions to be made during exploratory analysis (e.g. how should I proceed, what should I explore next and what questions should I ask), biases might interfere with the analysis. Systems have full knowledge of the data it contains and can perform unbiased computation on the data. A breadth-oriented exploration system can present useful information that users might have missed, and at the same time mitigate particular cognitive biases.

Understanding how breadth-oriented exploration alleviates biases is a crucial *future work*: Admittedly, the science of breadth-oriented exploration is still in its infancy. Several questions have to be answered before breadth-oriented exploration systems are widely adopted to mitigate biases during data exploration. For instance, what heuristics are resorted to when users explore data? While many heuristics are well-studied in psychology, how users apply them while navigating through data is less explored. Furthermore, what is the mechanism by which breadth-oriented exploration alleviates the biases caused by the use of these heuristics? What are the best strategies of mitigating biases by utilizing breadth-oriented exploration? We do not have answers to these questions but we believe that developing and evaluating more breadth-oriented systems is instrumental in answering these questions. Our design considerations can provide a starting point for system designers to explore the design space of breadth-oriented data exploration.

References

1. Chau DH, Kittur A, Hong JI, Faloutsos C (2011) Apolo: making sense of large network data by combining rich user interaction and machine learning. In: Proceedings of the SIGCHI conference on human factors in computing systems, ACM, pp 167–176
2. Dheeru D, Karra Taniskidou E (2017) UCI machine learning repository. http://archive.ics.uci.edu/ml/datasets/auto+mpg
3. Gotz D, Sun S, Cao N (2016) Adaptive contextualization: combating bias during high-dimensional visualization and data selection. In: Proceedings of the 21st international conference on intelligent user interfaces, ACM, pp 85–95

4. Kairam S, Riche NH, Drucker S, Fernandez R, Heer J (2015) Refinery: visual exploration of large, heterogeneous networks through associative browsing. Comput Graphics Forum, Wiley Online Library 34:301–310
5. Nickerson RS (1998) Confirmation bias: a ubiquitous phenomenon in many guises. Rev Gen Psych 2(2):175
6. O'malley AJ, Arbesman S, Steiger DM, Fowler JH, Christakis NA (2012) Egocentric social network structure, health, and pro-social behaviors in a national panel study of americans. PLoS One 7(5):e36,250
7. Perer A, Shneiderman B (2008) Systematic yet flexible discovery: Guiding domain experts through exploratory data analysis. In: Proceedings of the 13th international conference on intelligent user interfaces, ACM, pp 109–118
8. Sarvghad A, Tory M, Mahyar N (2017) Visualizing dimension coverage to support exploratory analysis. IEEE Trans Visualization Comput Graphics 23(1):21–30
9. Toms EG (2002) Information interaction: providing a framework for information architecture. J Assoc Inf Sci Technol 53(10):855–862
10. Tversky A, Kahneman D (1973) Availability: a heuristic for judging frequency and probability. Cognitive Psychology 5(2):207–232
11. Tversky A, Kahneman D (1974) Judgment under uncertainty: heuristics and biases. Science 185(4157):1124–1131
12. Van Ham F, Perer A (2009) ?search, show context, expand on demand?: Supporting large graph exploration with degree-of-interest. IEEE Trans Visualization Comput Graphics 15(6)
13. Willett W, Heer J, Agrawala M (2007) Scented widgets: improving navigation cues with embedded visualizations. IEEE Trans Visualization Comput Graphics 13(6):1129–1136
14. Wongsuphasawat K, Moritz D, Anand A, Mackinlay J, Howe B, Heer J (2016) Voyager: exploratory analysis via faceted browsing of visualization recommendations. IEEE Trans Visualization Comput Graphics 22(1):649–658

Chapter 12
A Visualization Approach to Addressing Reviewer Bias in Holistic College Admissions

Poorna Talkad Sukumar and Ronald Metoyer

12.1   Introduction

We set out to study the undergraduate admissions process at a highly-selective, private university in the United States that employs a holistic review process [1, 15]. Our goal was to design information visualization tools to aid in the process.

The university receives approximately 20,000 applications every year and has an acceptance rate of less than 20%. Every application is carefully reviewed by one to two reviewers and several factors are considered before making a recommendation. Additionally, the information contained in the applications are perused by the reviewers in largely text-based formats. Given these constraints, we and the admissions officers believe that the review process can benefit from visual tools tailor-made for this purpose. The tools can, for instance, ease the cognitive load experienced by the reviewers, reduce the time taken to review the applications, visualize the multivariate information contained in the applications and also visualize collections of applicants to help the reviewers reflect upon their decisions.

In order to design the visual tools, we proceeded to obtain a thorough understanding of what the process entails and the challenges faced by the reviewers by conducting observations and interviews. We were able to obtain a very-detailed picture of the multifaceted application-review process. One of the aspects that we had not considered previously, but stood out following our study, was the cognitive biases of the reviewers. These biases are understandable given the subjectivity inherent in holistic review processes.

Cognitive biases have been studied extensively and in various fields [6]. They are known to occur even in several everyday-scenarios of decision making [22].

P. T. Sukumar (✉) · R. Metoyer
University of Notre Dame, Notre Dame, IN, USA
e-mail: ptalkads@nd.edu

R. Metoyer
e-mail: rmetoyer@nd.edu

© Springer Nature Switzerland AG 2018
G. Ellis (ed.), *Cognitive Biases in Visualizations*,
https://doi.org/10.1007/978-3-319-95831-6_12

While not all of these biases have significant consequences, reviewer biases in the admissions process can have life-changing consequences. The admissions process is not only liable to the students but also to the respective institution's missions and goals.

We begin by describing related work followed by a brief characterization of the holistic review process. We then recapitulate the dual systems approach to thinking [13]—the theory that there are two mental systems, System 1 and System 2, from which many of our thinking processes and mistakes can be explained. We use the terminology of the dual systems throughout the chapter to communicate our ideas. We then provide arguments for how the review process can be classified as an example of a low-validity domain based on the conditions put forth by Kahneman and Klein [14]. This classification makes biases likely to occur during the process. We identify biases which may occur during holistic reviews and based on theories to counter them, suggest visualization strategies for their mitigation in the holistic-review process.

Our approach can prove useful in other similarly low-validity or insufficiently predictable domains, such as medical diagnostics and intelligence analysis, where judgments predominantly depend on expertise and subjective evaluations. The potential visualization strategies presented to mitigate the biases also contribute to the upcoming research area of addressing biases using visualization tools [16].

12.2 Related Work

12.2.1 Reasoning Heuristics and Cognitive Biases

We owe much of what we know today about reasoning heuristics and biases to the seminal work done by Tversky and Kahneman [13, 30]. They describe common heuristics employed in the reasoning process under uncertainty in their 1975 paper [30]. While these heuristics can be effective and lead to correct judgments in many situations, they can also sometimes lead to systematic errors or biases. These biases are not only exhibited by naive or inexperienced people but even experts are very likely to make such errors in low-validity environments. In this chapter, we borrow heavily from their work to make arguments for the potential biases identified in the the admissions review process and to suggest visualization strategies for bias-mitigation.

More recent research has found that emotions, known as the *affect* heuristic, can be a significant driving force in decision making and also that *priming* or suggestibility can occur in various indeterminable ways. For example, in a study of university admissions decisions by Simonsohn [24], incidental emotions were found to influence reviewer judgments. Simonsohn reported that applicants' non-academic attributes were weighted more heavily on sunnier days while their academic attributes were weighted more heavily on cloudy days. Additionally, judgments can also be influenced by the context, available choices, the framing of choices and similarity to previously encountered cases [12, 13].

The preconceptions and attitudes of people can also affect their decision making. Examples of the resulting biases include confirmation bias [18, 32] and myside bias, which is a type of confirmation bias where there is a tendency to favor evidence biased towards one's own opinions and attitudes [12].

Valdez et al. [31] provide a hierarchical framework for categorizing the various types of biases associated with visualizations. These categories can also be viewed as representing types of biases in general. The three broad categories include *perceptual* biases at the lowest level followed by *action* biases at the mid-level and *social* biases at the highest level. Perceptual biases are those that occur at the perceptual level, e.g., priming. Action biases refer to the reasoning biases occurring during decision making while social biases refer to those occurring on a social level including those influenced by culture and specific to individuals. We are predominantly interested in identifying and mitigating *action* biases as part of our study, i.e. the generalizable biases that can be identified through common reasoning heuristics employed in uncertain situations and are not specific to any individual or groups of individuals.

12.2.2   Using Visualizations to Mitigate Cognitive Biases

Using visual approaches for bias mitigation has shown promising results. Cook and Smallman evaluate a debiasing graphical interface, *JIGSAW*, developed to support analysis and mitigate confirmation bias in intelligence analysis [8]. They found that participants were less prone to confirmation bias using *JIGSAW* compared with a textual interface and this was because *JIGSAW* used a "recognition-centered" approach by making all the evidence, both supporting and conflicting with the hypothesis, constantly visible to the participants.

Visualization solutions have also been developed to facilitate Bayesian reasoning that enables users to overcome the errors made in solving and understanding conditional probabilities [17, 20, 27]. Sparklines, which are small graphics embedded within text [29], can be used to provide additional context and mitigate potential diagnostic errors in clinical data [21]. Visualization strategies have also been proposed to mitigate *planning fallacy*, based on potential causes found in prior literature [10], and perceptual biases occurring when font size is used to encode data [2].

12.3   Characterization of the Holistic Review Process

We conducted situated observations and interviews to learn about the holistic review process. Following approval from the Institutional Review Board, we conducted semi-structured interviews with 4 reviewers as well as observing them while they reviewed a sample application. The reviewers were asked to "think-aloud" while they reviewed an application. The individual interviews and observations each lasted between 30 min to an hour. We briefly describe the review process by distilling the key reviewer activities and tasks from the detailed data we collected.

In addition to reviewing applications, the reviewers perform other activities which contribute to their experience and knowledge which they utilize in assessing the applications. The reviewers are each assigned non-overlapping geographic areas and typically evaluate applications from these areas. They spend a considerable amount of time in visiting high schools in their areas as part of their recruitment activities and are usually familiar with the academic profiles and curriculum of the schools.

In a holistic review, every application is viewed in "context". Some of the factors that are considered by the reviewers include the applicant's high school grades and scores on standardized tests, applicant's high school, courses and opportunities offered by the school, other student profiles in the school, relation between the courses taken by the applicant and his or her "major" preference(s), family background, adversities faced, trends in grades, explanations for dips in performance, non-academic accomplishments of the applicant, leadership qualities, special talents, and qualities of the applicant gleaned from letters of recommendation and the applicant's essays (e.g., "risk-taker", "perfectionist", and "vulnerable"). Additionally, the reviewers are also guided by the university's missions and policies, such as diversity and upholding the values and personal qualities advocated by the institution. While we observed some overlap in how the reviewers considered the different application attributes, many of the judgments depended on individual reviewers' experiences and intuitions.

We also observed roughly 2 hours of a mock committee meeting session. Committee meetings take place after the individual reviews of applications. Reviewers present selected applications they reviewed during these meetings and the committee discusses these applications and makes recommendations on undecided applications.

12.4 System 1 and System 2

Kahneman, in *Thinking Fast and Slow* [13], describes the dual systems approach to thinking comprised of a fast, automatic, and intuitive system which he calls System 1 and a slow, reflective, and conscious system which he calls System 2. Additionally, System 1 is responsible for making associative connections in memory and has a tendency to find coherence and attach causal explanations to events, even where there is none. System 2, on the other hand, is associated with problem solving, subjective experiences, and self control; it is also known to be lazy and often unthinkingly endorses the suggestions made by System 1. These systems, with their distinct functionality, provide us with a useful terminology to understand and discuss our various judgment and decision-making processes. We refer to these systems in our discussions throughout the rest of this chapter.

12.5   Accuracy of Expert Intuition in Holistic Reviews

Both expertise and heuristics contribute to the intuitive thinking of System 1 [13]. In numerous domains, experts instinctively recognize the problem or figure out solutions without employing the conscious thinking of System 2, e.g., chess masters being able to reproduce a chess position after viewing the board for only a few seconds [7]. Not only is this possible due to their extensive practice and profound familiarity with the domain, but it is also necessary for the domain to be relatively regular or predictable and afford users ample opportunity to learn the numerous variations through practice and efficient feedback [13, 14].

Heuristics, on the other hand, essentially function by means of "substitution", i.e. to answer a given question, we often substitute it by answering a simpler question [13, 30]. There are a few known patterns in which the substitutions can occur and these can result in different types of cognitive biases. We discuss these biases in Sect. 12.6. We use heuristics in our judgments and decisions in everyday life. They are apt for most situations since invoking the thoughtful processes of System 2 is usually unnecessary and also impractical due to time and processing constraints of the human brain. However, there are situations, especially those involving uncertainty, where these heuristics can lead to systematic errors or biases.

To assess the reliability of expert intuitions in any given domain, the conditions which define the *validity* of the domain must be considered: the regularity or causal structure of the domain and the opportunities provided by the domain to learn the regularities and develop accurate intuitions [14]. The game of chess, for example, has high validity. It is sufficiently regular with defined moves and outcomes and players can learn the gamut of possible moves through adequate practice and develop *reliable* intuitions [7, 14].

Validity should not be confused with uncertainty as uncertainty can exist even in high-validity domains [14]. The validity of a domain tells us if the experts can develop accurate intuitions in the domain which, even in the presence of uncertainty, can improve the chances of success. If a domain exhibits low validity, it means that the intuitive thinking of System 1 will most-likely employ heuristics as opposed to expertise which can lead to systematic errors or biases in certain scenarios.

Hence to determine the accuracy of the expert intuition of reviewers and the validity of the holistic-review process, we need to consider the following questions:

1. Is the domain of holistic reviews sufficiently regular and predictable?
2. Does the review process afford reviewers sufficient opportunity to learn all the possible scenarios and variations in applications, and the most appropriate recommendations to make for them through timely feedback?

It can be argued that the holistic-review process is not sufficiently regular because not only is there a substantial variation among the applications received by the university but the most appropriate recommendation to make for each is also not obvious.

The main measure of the reviewers' competence is how well the selected students perform in their 4 years at the university and how their overall performances conform to the expectations and standards of the university. Not only is this feedback much delayed but it also does not provide any information on what the results would be if a different set of students had been selected.

We also gathered from our study that the work experience of the reviewers ranged from 1 to more than 30 years. While more-experienced reviewers may have had the opportunity to evaluate many types of applications and hone their evaluation methods over the years, the less-experienced reviewers may only be familiar with a small subset of applications and their reviewing procedures may not be sufficiently insightful.

Even if we optimistically consider all the applications the university receives to broadly fall into a certain number of types and the review process to afford reviewers to become familiar with each of these types, the reviewers may still be unable to make optimal recommendations. This is because of the late and insufficient feedback received as well as the constantly-changing admissions scenario in the United States. For instance, the number of applications received by universities are generally increasing and there is more diversity in the the demographic composition of applicants forcing universities to adapt to these changes and alter their decision-making strategies.

Hence it can be said that the holistic-review process is a domain with low validity or predictability in which the reviewers can be susceptible to cognitive biases.

12.6   Possible Reviewer Biases

In many domains, including the admissions review process, medical diagnostics and law, it is non-trivial or not possible to define what the most optimal judgments or profitable behavior should be; the adopted approach is to instead identify the biases or deviations from what is considered rational behavior [25, 30]. We too follow this approach and rather than focus on what is ideal, we determine the potentially non-optimal ways in which reviewers can judge the applicant attributes based on the data we collected from our study.

We take a user-centered and task-centered approach to deconstruct the types of biases in reasoning in the holistic review process. As a result of our interviews and observations, we were able to obtain an exhaustive list of the tasks the reviewers perform while reviewing every application. These tasks mainly consist of the various application attributes they examine and their thought processes contributing to the judgments they make concerning each attribute and overall. To identify potential reviewer biases, we matched these task descriptions with the tasks associated with the heuristics of reasoning under uncertainty and the known biases in decision making [4, 6, 13, 18, 30].

For example, the *representativeness* heuristic [30] is generally associated with tasks or questions that ascertain how similar A is to B or how representative A is of B. We found such tasks performed during the application reviews and matched them with this heuristic. We provide a few such examples below.

12.6.1   *Coherence, Causal Associations, and Narrative Fallacy*

System 1 is inclined towards finding coherence in the information presented and attributes causal explanations to events which may be random or chance events [13]. It is also associated with "narrative fallacy" which is making sense of past events, for example, to come up with "tidy" success stories and undermining the value of luck and chance [13]. We observed in our study that the reviewers make many such connections while reviewing applications. For example, in assessing the grade trends of the student, if any grade(s) stands out, they try to reason why and they often link aspects observed in the student's essays with their performance. They may also exhibit narrative fallacy when discussing applications during committee meetings. These reflect System 1's tendency to find causal interpretations and disregard luck in assigning value to one's talent and personality.

These System 1 traits often increase the confidence in our judgments and provide us with the *illusion of validity*. However, confidence in judgments should not be considered a measure of their accuracy. Kahneman states that this overconfidence and optimism in our judgments may very well be the most significant of the cognitive biases [13]. Being unsure of our judgments and aware of uncertainty and the role played by chance is a more rational means of decision making. The irony, however, is that confidence in individuals, especially in professionals such as clinicians, lawyers and even admissions reviewers, is generally perceived and regarded very highly and a quality associated with successful people.

12.6.2   *Anchoring as Adjustment*

Anchoring occurs when a piece of information encountered is considered consciously or unconsciously in making a subsequent estimation [13]. It can occur in the form of priming where System 1 unknowingly falls prey to the suggestibility, or as adjustment where there is a deliberate but insufficient switch made from the anchor involving System 2. Since priming can occur in several indeterminable ways, we are more interested in anchoring as adjustment.

Anchoring can occur both between applications and within a single application during the review process in admissions. The application(s) reviewed prior to the current one, may act as the anchor and alter the reviewer's expectations preventing them from reviewing each application with a clean slate.

We observed in our study that the reviewers generally reviewed the academic scores of an applicant first before evaluating other attributes. Hence the academic performance of an applicant can be thought of as the *anchor* or starting point proceeding from which other aspects are estimated by *adjustment*. This makes possible the biases of *overestimating conjunctive events* and *underestimating disjunctive events* [13, 30]. For example, if a student has excelled academically, then the reviewers may be more optimistic regarding the student's success in other aspects (conjunctive events) and underestimate the chances of finding an aspect in which the student has performed poorly (disjunctive event).

12.6.3 The Halo Effect

The halo effect is exhibited when one tends to like (or dislike) *everything* about a person based on gathering only a few traits of the person [13]. The impressions are usually anchored by the first piece of information they come across and are extrapolated to even characteristics that have not been observed. The halo effect can be described as the result of combining emotion on judgments (known as the "affect heuristic"), System 1's need for coherence, and the anchoring effect.

The halo effect can potentially occur both during individual application reviews and while presenting applications during committee meetings. Reviewers may form strong opinions about the applicants based on the order and information encountered in the applications and this can not only affect the recommendations they make but can, in turn, lead to them influencing the committee members when presenting the applications.

12.6.4 Confirmation Bias

Confirmation bias [13, 18] is one of the manifestations of the associative memory of System 1 which finds confirming (as opposed to disconfirming) evidence of any given statement or scenario.

In the applicant review process, reviewers may be inclined to find or favor evidence in the applications that confirms their judgments about an applicant and ignore or disfavor evidence that disconfirms their judgments. Additionally, it has been found that people exhibit confirmation bias even when recalling information [6] and hence reviewers may also be subject to this bias while presenting their applications during the committee meetings.

12.6.5 Availability

The availability heuristic substitutes the judgements concerning an event with the ease and associated emotion (i.e. affect heuristic) with which instances pertaining to the event come to mind [13]. It plays a role in the review process when the evaluation of the application attributes is affected by the ease with which instances, or information related to the attributes, can be recalled. For example, reviewers tend to better remember the students they interacted with during their high-school-visits and immediately recognize the student from his or her application. This will lead to the bias caused by the *retrievability of instances* [30]. Reviewers are generally very familiar with the high schools in their region and hence tend to recall a lot of information about a school from memory when reviewing an applicant from the school. The information recalled may be incomplete or biased.

Reviewers are likely to remember the attributes of an applicant that stand out and this may influence their judgments regarding other aspects of the applicant. For example, if a reviewer perceives a student as coming from a privileged background, the reviewer may use this information to form opinions on other aspects of the student such as having ample opportunities to pursue certain activities or start non-profit organizations. Such instances can lead to the biases due to the *retrievability of instances* as well as *illusory correlation* [30] wherein strongly-associated events are thought to frequently co-occur.

Reviewers present applications during committee meetings and use their notes and summaries recorded previously to recall the respective applications. Since committee meetings take place at a later time, the biases due to the *retrievability of instances* [30] may play a role when reviewers discuss their applications.

12.6.6   Representativeness

The *representativeness* heuristic refers to System 1's use of similarity and stereotype information rather than base rates to infer probabilities in many situations. In holistic reviews, this heuristic plays a role when assessing how representative an applicant's attributes are of an existing or predetermined set.

In assessing an applicant's fit to the university, reviewers assess how representative an applicant is of the values and key aspects advocated by the institution by considering attributes such as the overlap between the interests of the student and those of the institution and judging the applicant's personal qualities. The biases due to *insensitivity to predictability* and *the illusion of validity* [30] can occur in this case when the attributes considered are not actually predictable of the student's fit to the university but nevertheless, the reviewer is very confident of his or her judgments.

Reviewers assess how *representative* an applicant is to the group of students from the same school who were admitted to the university in the previous years. In doing so, the biases due to *insensitivity to prior probability of outcomes* and *insensitivity to sample size* [30] can occur when the reviewers mostly consider the application attributes such as the *average* Grade Point Average, i.e. GPA (when the sample size is small, the sample statistic can differ significantly from the population parameter), but do not adequately consider potentially vital information, such as the GPA ranges and percentages of students from this school admitted to the university in the past years.

12.6.7 The Avoidance of Cognitive Dissonance

Cognitive dissonance refers to the conflict experienced when opposing pieces of evidence are people feel uneasy and to placate the conflict, the inconsistency is usually resolved by weighting one piece of evidence more favorably than the other [4, 5].

Reviewers may find themselves in situations wherein they are both impressed by certain aspects of the applicant and not very impressed by certain other aspects. In order to make a decision, they will have to resolve this dissonance by *adjusting* their beliefs. As a result, they may overestimate the achievements and underestimate the setbacks and recommend to admit the student or vice versa.

12.6.8 Time-Induced and Stress-Induced Biases

Switching between tasks involving System 2 and performing tasks requiring several ideas to be retained in mind simultaneously can be taxing, especially under time pressure [13]. The negative effects of time pressure and stress on human judgment and decision making have been studied extensively [26]. These effects include a reduced search for information, making defensive choices, reinforcing the choices made, a tendency to process information according to their perceived priority, giving more importance to negative information and forgetting crucial data [34].

Time is a critical factor in the application review process in admissions. One of the main challenges that the reviewers face is the sheer volume of the applications and the limited time to review them which often results in the reviewers putting in extra hours of work. In our cohort of reviewers, each review between 1000 and 2000 applications in a time period of roughly 2 months and typically take about 15 min to review each application. This time not only includes their thought processes and consideration of the different application attributes but also the time taken to write notes and enter ratings. Needless to say, the reviewers are subject to considerable time pressure and stress while reviewing the applications and hence the time-induced biases mentioned above are likely to occur during the process.

12.7 Proposed Visualization Strategies to Mitigate Biases

We propose the following visualization strategies for the holistic review process based on the potential biases and reviewer challenges identified in our study. These strategies draw from solutions in the cognitive-bias literature to counter the fallacies of System 1 as well as from the visualization literature on tools and techniques used to aid sensemaking. These strategies can be adapted for use in other domains also employing cognitive-intensive and subjective assessments, such as medical diagnostics and intelligence analysis.

12.7.1   Easing Cognitive Load

The applications are currently read by reviewers in largely text-based formats. Designing visual tools for the application review process may not only communicate certain types of information more easily but could also potentially ease the cognitive load of the reviewers and mitigate some of the time-induced and stress-induced biases [11].

Visual representations of the application attributes along with the corresponding additional statistics considered, such as the GPA ranges and percentages of students admitted in the previous years from the high school, can aid in reviewers' decision making as well as mitigate the availability bias by presenting all the information needed including those easily recalled and not recalled to make the recommendation [16].

For more-subjective evaluations, such as assessing "fit" and the leadership and non-academic skills of the applicant, it may be helpful for the reviewers to collectively discuss and formulate the various definitions of what they look for or what the university represents, and rank them from the most important to the least important. This ordinal/nominal information can be presented visually alongside the attributes considered in the application. This can not only help reduce the bias but also standardize the task assessments.

12.7.2   Supporting Sensemaking

The holistic review process involves cognitive-intensive practices analogous to sensemaking. Several pieces of information are considered and carefully reviewed to make recommendations. Hence visualization tools derived from the sensemaking domains, such as efficient note-taking and snapshot tools [3, 23, 33], can be beneficial for the reviewers to better record their rationales for the recommendations made. These can not only aid in saving the intermediate judgments and tie them to data evidence, but they can help the reviewers to better recall applications during the committee meetings which occur at a later time, thereby mitigating narrative fallacy, availability and confirmation biases.

Additionally, many types of biases including confirmation bias, anchoring bias and biases due to illusory correlation and cognitive dissonance can be mitigated by presenting alternative visual representations of the application attributes to enable the reviewers to consider other possible interpretations [35]. This may not only enable them to weigh the attributes more suitably but also permits more introspective evaluations.

12.7.3   Decorrelating Error

Biases can not only occur at various points in the review process but can also accrue. Our study findings indicate that the final recommendations reviewers make on applications can be viewed as combinations of smaller judgments made on several individual aspects of the application. Hence biases occurring in the judgment of one aspect can propagate to the judgments made on other related aspects. Decorrelating error is a suggested method to counter the Halo effect and anchoring bias [13]. It would involve breaking down applications into the various attributes and reviewing them independently as opposed to reviewing a student's application as a whole. This can prevent the judgement errors made with respect to individual attributes from becoming correlated when an application is reviewed in its entierly. This will also prevent the reviewers from making spurious causal interpretations in an application.

The visualization interface for reviewing applications can present the attributes independently to the reviewers and record their ratings and assessments for a respective attribute using the visual note-taking tools. The overall recommendation for an applicant can be made by putting together these individual attribute-reviews.

This approach may diminish the reviewers' confidence in their recommendations but being less certain is reflective of a more rational approach to decision making [13].

12.7.4   Mobilizing System 2

Most of the biases described occur due to the intuitive suggestions of System 1 which are unthinkingly endorsed by the usually lazy System 2. Hence mobilizing System 2 may help rethink and disregard the suggestions of System 1 [13]. For example, in the availability heuristic, System 1 usually goes by the *ease* with which instances come to mind but an active System 2 would also focus on the *content* of the instances retrieved to make better-informed decisions. Additionally, it has been suggested that some individuals are more prone to biases than others because they exhibit a certain unwillingness to involve the conscious thinking of System 2 in many situations, or in other words, their System 2, while not necessarily inept, is lazier [13, 25]. Hence activating System 2 can potentially help in mitigating many of the aforementioned biases.

One of the ways in which System 2 can be activated is by inducing *cognitive strain* in the visualization interface; for example, by making the text and visualization harder to read [13]. This strategy goes against the general principles of interface and visualization design [19, 28] which advocate reducing the cognitive load of users. While these principles generally focus on the most effective presentation of the data, visualization strategies for bias mitigation may require employing unconventional and even conflicting designs to mobilize System 2. Similarly, Correll and Gleicher make the case for how embellishments and distortions in visualizations can be beneficial to foster knowledge and decision making in certain scenarios [9].

12.7.5 Combining Formulas with Intuition

In many professional domains including holistic reviews in admissions, medical diagnostics, court cases, and intelligence analysis, human assessments are considered indispensable and there is skepticism about the use of formulas to make decisions. However, research has shown that simple rules and formulas that involve assigning values to known predictive variables perform more accurately than expert evaluations, especially in low-validity domains [13].

We found in our study that reviewers spend considerable time in visiting the high schools to gather information and meet potential students. We can see that these experiences and knowledge of reviewers are advantageous to the application-review process since they enable them to be familiar with the contexts of the applicants. However, given the low-validity of the domain and the possibility of biases occurring during the review process, it might be beneficial to combine reviewer judgments with formula-based or rating-based approaches.

Kahneman describes how intuition works more accurately after objectively analyzing and scoring the variables involved [13]. Similarly, reviewers can be asked to consider each application attribute independently (as described under "decorrelating error") and assign a score to it based on their judgments. These separate scores can then be viewed collectively for every applicant and based on these, a final recommendation can be made by the reviewer. This method can not only yield results with more accuracy than a purely holistic, subjective assessment approach but they also involve a more consistent procedure without requiring unnecessary complex thinking [13].

12.8 Conclusion

Numerous universities throughout the country employ holistic reviews. This approach provides an equitable means to review student applications by contextualizing their performance in light of the opportunities they have been presented and supports the shaping of a class that is representative of the respective university's mission and goals. However, reviewer biases are a probable occurrence during the process because it involves subjective assessments and inadequate opportunities for fostering the right intuitions pertaining to making admissions decisions. These reviewer biases, in turn, can have direct consequences for the applicants and also, indirect social, economic and political ramifications for the country as a whole.

We have interweaved the theoretical concepts surrounding cognitive biases into (i) our understanding of the holistic-review process in admissions based on data collected from ethnographic studies and (ii) formulating visualization strategies to mitigate the potential biases in the review process. This chapter highlights the probable occurrence of biases in consequential, expertise-dependent, low-validity domains that can significantly affect the decisions made.

Our proposed visualization strategies not only suggest significant changes to the current practice of the review process but also hint at employing unconventional visualization designs in mitigating the biases, such as 'cognitive strain' (as outlined in sect. 12.7.4).

Acknowledgements We wish to thank the admissions officers at the university where we conducted our studies for their cooperation and participation in our research study. This material is based upon work supported by the National Science Foundation under Grant No. 1816620.

References

1. (2002) Best practices in admissions decisions. A Report on the Third College Board Conference on Admission Models. College Entrance Examination Board
2. Alexander EC, Chang CC, Shimabukuro M, Franconeri S, Collins C, Gleicher M (2017) Perceptual biases in font size as a data encoding. IEEE Trans Visual Comput Graph 3(9):1667–1676
3. Andrews C, Endert A, North C (2010) Space to think: large high-resolution displays for sensemaking. In: Proceedings of the SIGCHI conference on human factors in computing systems, ACM, pp 55–64
4. Aronson E (1969) The theory of cognitive dissonance: a current perspective. Adv Expe Soc Psychol 4:1–34
5. Aronson E, Mills J (1959) The effect of severity of initiation on liking for a group. J Abnorm Soc Psychol 59(2):177
6. Burke AS (2005) Improving prosecutorial decision making: some lessons of cognitive science. William & Mary Law Rev 47:1587
7. Chase WG, Simon HA (1973) Perception in chess. Cogn Psychol 4(1):55–81
8. Cook MB, Smallman HS (2008) Human factors of the confirmation bias in intelligence analysis: decision support from graphical evidence landscapes. Hum Factors 50(5):745–754
9. Correll M, Gleicher M (2014) Bad for data, good for the brain: knowledge-first axioms for visualization design. In: IEEE VIS 2014
10. Dragicevic P, Jansen Y (2014) Visualization-mediated alleviation of the planning fallacy. In: IEEE VIS 2014
11. Fekete JD, Van Wijk JJ, Stasko JT, North C (2008) The value of information visualization. In: Information visualization. Springer, pp 1–18
12. Goldstein EB (2014) Cognitive psychology: connecting mind, research and everyday experience. Nelson Education, Scarborough
13. Kahneman D (2011) Thinking, fast and slow. Macmillan, New York
14. Kahneman D, Klein G (2009) Conditions for intuitive expertise: a failure to disagree. Am Psychol 64(6):515
15. Lucido JA (2014) How admission decisions get made. In: Handbook of strategic enrollment management pp 147–173
16. MacEachren AM (2015) Visual analytics and uncertainty: its not about the data
17. Micallef L, Dragicevic P, Fekete JD (2012) Assessing the effect of visualizations on bayesian reasoning through crowdsourcing. IEEE Trans Visual Comput Graph 18(12):2536–2545
18. Nickerson RS (1998) Confirmation bias: a ubiquitous phenomenon in many guises. Rev Gen Psychol 2(2):175
19. Norman D (2013) The design of everyday things: revised and expanded edition. Basic Books (AZ), New York
20. Ottley A, Peck EM, Harrison LT, Afergan D, Ziemkiewicz C, Taylor HA, Han PK, Chang R (2016) Improving Bayesian reasoning: the effects of phrasing, visualization, and spatial ability. IEEE Trans Visual Comput Graph 22(1):529–538

21. Radecki RP, Medow MA (2007) Cognitive debiasing through sparklines in clinical data displays. AMIA Annual Symposium Proceedings 11:1085
22. Ross HJ (2014) Everyday bias: identifying and navigating unconscious judgments in our daily lives. Rowman & Littlefield, Minneapolis
23. Shrinivasan YB, van Wijk JJ (2008) Supporting the analytical reasoning process in information visualization. In: Proceedings of the SIGCHI conference on human factors in computing systems, ACM, pp 1237–1246
24. Simonsohn U (2007) Clouds make nerds look good: field evidence of the impact of incidental factors on decision making. J Behav Decis Making 20(2):143–152
25. Stanovich K (2011) Rationality and the reflective mind. Oxford University Press, New York
26. Svenson O, Maule AJ (1993) Time pressure and stress in human judgment and decision making. Plenum Press, New York
27. Tsai J, Miller S, Kirlik A (2011) Interactive visualizations to improve Bayesian reasoning. In: Proceedings of the human factors and ergonomics society annual meeting, vol 55. SAGE Publications, Los Angeles, CA, pp 385–389
28. Tufte E, Graves-Morris P (2014) The visual display of quantitative information (1983)
29. Tufte ER (2006) Beautiful evidence. Graphics Press, New York
30. Tversky A, Kahneman D (1975) Judgment under uncertainty: heuristics and biases. Utility, probability, and human decision making. Springer, The Netherlands, pp 141–162
31. Valdez AC, Ziefle M, Sedlmair M (2017) A framework for studying biases in visualization research. In: IEEE VIS 2017
32. Wason PC (1960) On the failure to eliminate hypotheses in a conceptual task. Q J Exp Psychol 12(3):129–140
33. Wright W, Schroh D, Proulx P, Skaburskis A, Cort B (2006) The Sandbox for analysis: concepts and methods. In: Proceedings of the SIGCHI conference on Human Factors in computing systems, ACM, pp 801–810
34. Zakay D (1993) The impact of time perception processes on decision making under time stress. In: Time pressure and stress in human judgment and decision making. Springer, pp 59–72
35. Zuk T, Carpendale S (2007) Visualization of uncertainty and reasoning. In: Smart graphics. Springer, pp 164–177

Chapter 13
Cognitive Biases in Visual Analytics—A Critical Reflection

Margit Pohl

13.1 Introduction

There is ample evidence that humans tend to commit cognitive biases under some circumstances [9]. Humans also have difficulties with logical thinking and reasoning [8]. Nevertheless, research addressing these issues has also been criticized [17].

It has been argued that the experiments that substantiate this research do not reflect realistic problem-solving processes. They often use puzzle problems or highlight abstract logical problems that are fairly artificial. Puzzle problems are specifically designed to exclude context and background knowledge. There is a good reason for doing this from a methodological point of view because context and background knowledge are confounding variables that have to be kept constant so that experiments will yield reliable results. In psychology, there is often a trade-off between methodological rigor and ecological validity. Puzzle problems also do not require extended problem-solving processes. There are clear solutions and the paths to these solutions are unambiguous, sometimes even algorithmic processes.

Visual analytics, in general, supports exploratory processes in ill-structured domains (e.g. in medicine, intelligence analysis, finance). In ill-structured domains, there are neither clear-cut solution methods nor easily identifiable solutions. In addition, visual analytics works with very large amounts of data, much of which is unnecessary and distracting. Most of the domains where visual analytics is applicable require a great deal of expertise as a foundation for successful problem-solving processes. Problem-solving in such areas often requires a heuristic approach rather than formal logic or the ability to solve puzzle problems. Such heuristic approaches are also based on rational, goal-directed thinking—there are other forms of rational thinking beyond formal logic or solving puzzle problems [4]. It is an open question

M. Pohl (✉)
Institute of Design and Assessment of Technology, Vienna University of Technology, Vienna, Austria
e-mail: margit@igw.tuwien.ac.at

© Springer Nature Switzerland AG 2018
G. Ellis (ed.), *Cognitive Biases in Visualizations*,
https://doi.org/10.1007/978-3-319-95831-6_13

as to what extent research on cognitive biases is applicable in visual analytics. This chapter reviews research from cognitive psychology and tries to clarify some of the open issues using an example from intelligence analysis.

13.2 Puzzle Problem Approach Versus Everyday Thinking and Reasoning

Kahneman [9] is one of the most well-known representatives of a dual process theory of thinking and reasoning. He argues that there is System 1 that is fast but tends to be error-prone and System 2 that is slow and relies on logical thinking and therefore leads to more correct results. System 1 decisions tend to be influenced by cognitive biases because they rely on gut feeling more than on logical reasoning. This view has been criticized by some authors.

Evans [4], for example, argues that the concept of cognitive biases is based on a conception distinguishing between rational and irrational decision-making. This conception presupposes some normative framework which enables researchers to differentiate between decision-making processes conforming to the norm and others that do not. In general, formal logic or probability theory are defined as the normative standard and behavior deviating from this is seen as irrational. Evans, however, points out that human reasoning is, by design, pragmatic rather than logical. Therefore, an assessment of reasoning and decision-making processes based exclusively on the norm of logical thinking might distract from the actual mechanisms governing these processes.

In addition, it should be pointed out that it is not clear how rationality can be defined. Traditionally, it is equated with areas like formal logic, propositional logic or probability theory. Woll [17] argues that there is no one model of logic or probability, and that these concepts are constantly changing. This makes it fairly difficult to define a normative foundation for identifying cognitive biases. In addition, some researchers have developed alternative concepts of rationality, such as the concept of ecological rationality by Gigerenzer [7]. Ecological rationality studies the mind in its relationship to its environment and how humans adapt to the affordances of this environment during reasoning processes. Gigerenzer argues that reasoning processes evolved as a consequence of the humans' interaction with the world around them. This is a development comparable to the evolution of the human senses that are adapted to the specific nature of light, sound and other stimuli surrounding human beings. In the context of this approach, heuristics and gut feeling are not seen as irrational, subjective and error-prone, but as very complex, fast and effective methods to cope with the necessities of our daily lives. This approach deviates significantly from Kahneman's [9] ideas that emphasize the flawed and imperfect character of heuristics. Heuristics in the model developed by Gigerenzer are comparable to well-defined algorithms and are often domain-specific. This means that Gigerenzer's model takes background knowledge into account.

This discussion is especially relevant for visual analytics because interaction with such systems is seen as exploratory sensemaking processes rather than as drawing logical conclusions. Visualizations support looking at data from different points of view and formulation of competing hypotheses. A concept of thinking and reasoning based on some normative framework might be too restrictive to model these processes in the context of visual analytics.

Given the exploratory nature of interaction processes with visual analytics systems, it is no surprise that they are often seen as sensemaking processes producing insights rather than information. There are several theories that have influenced this approach. One of the most influential is Klein's sensemaking model. Klein et al. [10, 11] developed the data/frame model, which is based on natural decision-making (NDM). The main focus of NDM is to analyze decision-making by domain experts in complex situations with the goal of modeling sensemaking activities in naturalistic settings. Klein and his co-authors also assume that realistic situations, expertise in a given domain and use of pragmatic and flexible strategies play an important role. They posit that people develop schematic representations called frames. These frames can be elaborated, questioned or rejected. This model has been applied successfully to explain interaction with visual analytics systems because it reflects the exploratory processes going on while users are gaining insights and the importance of domain-specific knowledge.

Another model reflecting the exploratory sensemaking processes of users of visual analytics systems is the knowledge generation model of Sacha et al. [16]. In this model, there are three different loops of cognitive activity: the knowledge generation loop, the verification loop and the exploration loop. This model emphasizes the importance of uncertainty about the data and the resulting problems in decision-making processes. Incorrect or corrupt data also require complex sensemaking processes that go beyond logical thinking. Users need efficient heuristics based on domain knowledge to cope with such problems.

Norman [14] argues that for the process of medical diagnosis there is evidence that the main source of error is lack of knowledge, not cognitive bias. This is overlooked in a discussion focusing on cognitive biases. In addition, he points out that extensive literature on the distinction between experts and novices shows that the true expert relies more on intuitive reasoning of the System 1 type, while novices apply deliberative rule-based methods.

Fiedler and von Sydow [6] provide an overview of research concerning cognitive biases based on Kahneman's basic assumptions. They argue that this type of research is too vague to serve as an underlying theory to explain how cognitive biases develop. It is not clarified which cognitive processes are executed when such biases occur. Many of the assumptions underlying the research in cognitive biases are not experimentally manipulated to test them systematically (e.g. the availability of information in the availability heuristics). Nevertheless, Fiedler and Sydow [6] point out that, despite its weaknesses, the research on cognitive biases has given rise to an extensive research program into thinking and reasoning processes that has clarified other interesting issues.

Woll [17] provides an overview of the discussions about cognitive biases. He argues that there are methodological issues with the approach of Kahneman. Several researchers pointed out that the experiments conducted by Kahneman are fairly artificial and designed in a way to generate cognitive biases. Study participants have to make decisions based on scant information without any context. In contrast to that, decisions in everyday situations are usually based on redundant information. People often possess a considerable amount of background knowledge and decision-making is a long-term process with several feedback loops. Woll doubts that the results from laboratory research on cognitive biases is applicable to such situations. In everyday thinking and reasoning people generally do not apply formal decision-making processes, but rather adapt their strategies to the problem at hand and use these strategies very flexibly. Context and background knowledge play an important role.

I want to illustrate this line of argument using a well-known example from cognitive psychology—Watson's selection task. This task is one of the best researched tasks in cognitive psychology (for an overview see Eysenck and Keane [5]). In this task, study participants see four cards, two of which show letters and two numbers. These cards also have numbers or letters on the reverse of the card that is not visible. Participants then get a rule: If there is a vowel on one side of a card, then there is an even number on the other side of the card. The tasks the participants have to solve is which cards they have to turn around to find whether this rule applies or not. In this abstract form, this task is fairly difficult. It can be shown, however, that this task gets much easier when it is embedded in a concrete context (e.g. if a letter is sealed it has a 5d stamp on it instead of the rule concerning vowels and even numbers). There has been much controversial discussion about this phenomenon, and it is not entirely clear how the influence of context information can be explained, but in general, this seems to be a fairly stable result. The model of pragmatic reasoning schemata has been proposed to explain this phenomenon [5]. This model assumes that there are rules that apply to certain classes of situations. In this context, knowledge about the world is essential.

13.3 Bias Mitigation Strategies

Several different cognitive bias mitigation strategies have been discussed in the literature. Nussbaumer et al. [15] especially discuss the following bias mitigation strategies: providing different views of the data to change the perspective; providing information about the uncertainty of the data; computerized critique questions; explicit prompts to rethink one's own hypotheses; discussion of hypotheses with peers; visualization of multiple hypotheses. Kretz et al. [12, 13] especially study cognitive bias mitigation strategies in the context of intelligence analysis. They point out that there is still a lack of systematic empirical studies about the efficiency of bias mitigation strategies, especially in realistic contexts. They tested several different bias mitigation strategies and found that some of them are more efficient than others. In addition, it depends on the context which of the bias mitigation strategies are more efficient than others.

Bias mitigation strategies can have beneficial effects on the quality of decision-making. In an evaluation of a system for intelligence analysts, we could show that providing two different visualizations of one and the same data set motivated some users to adopt a verification strategy to make sure that their results were also supported by the second visualization [3]. However, many of these strategies require the users to spend additional time and effort. Given the time constraints under which, for example, intelligence analysts operate they will be reluctant to adopt such strategies.

Intelligence analysts have a considerable amount of background knowledge. They are able to use context to arrive at valid results. Based on the discussion about the importance of context and background knowledge it might be argued that this might help analysts to avoid cognitive biases. As described above, there is some evidence indicating that cognitive biases especially occur in laboratory situations where study participants are only provided with a minimal amount of information. As a consequence, it might be argued that by providing context and activating background knowledge cognitive biases can be avoided.

I want to illustrate this argument with an example from our work with intelligence analysts. Intelligence analysts often work with network visualizations of co-offenders, that is offenders who commit crimes together (Fig. 13.1). This system is described in more detail in Doppler-Haider et al. [3]. Figure 13.1 just shows a node-link diagram consisting of icons for the offenders and links to indicate that these offenders committed a crime together. In addition, there is some information about the temporal development of this cooperation and the seriousness of the crimes (which is indicated by the weight of the line) that were committed. The goal of this visualization is to support intelligence analysts in the investigation of co-offender networks, for example, whether the criminal activities of a specific network increases or not or whether the types of crimes committed by such a network changes over time. Nevertheless, for a real task of an intelligence analyst the information shown here is still insufficient. Analysts need detailed information about the specific crimes that were committed to be able to assess their development and the specific contacts that the co-offenders have [1]. Such systems should, for example, provide specific information about the offenders and the crimes they committed. This information should be easily accessible from the node-link diagram shown in Fig. 13.1. On the other hand, experienced analysts already possess a lot of background information about the criminal activity in their area. This kind of information should also be activated. The information system should be designed in a way that analysts can easily combine their own background knowledge with the new knowledge provided by the system.

From informal observation of analysts, we know that they need to interact with much of the data to get a comprehensive overview and a feel for the data. Nevertheless, in practice, it is not straightforward to decide which kind of data to provide to analysts, when to provide these data and how to activate their background knowledge.

Dimara et al. [2] conducted an experiment to find out whether adding context can increase the accuracy of task solutions when participants work with visualizations.

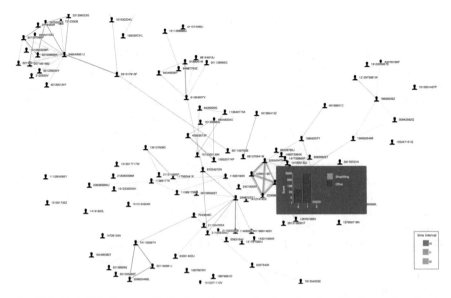

Fig. 13.1 Node-link diagram of offenders who commit crimes together. The three colors denote three different years. When co-offenders are linked by more than one line, this indicates that they worked together in several years. The thickness of the line indicates the weight, that is, the serious-ness of the crime according to the priorities of police forces

They found that this is not the case. They found, however, that it does increase confidence and the user experience. This is a result that indicates that adding context is not a straightforward strategy. I would like to point out, however, that this experiment used crowdsourcing and fairly brief narratives to provide context. From the point of view of everyday thinking and reasoning, it might be argued that this is still a fairly artificial situation. Nevertheless, the study indicates that providing context has to be designed with care in order to be successful.

13.4 Conclusion

Research on cognitive biases is a very interesting and challenging area of research. It is obvious that cognitive biases occur under certain circumstances. Nevertheless, it is not entirely clear how relevant this research is for visual analytics. Visual analytics operates in ill-structured domains. Therefore, it is often difficult to apply highly for-mal methods of reasoning (as, for example, formal logic). Cognitive bias research is usually based on such formal methods as a normative foundation. This is one reason why it might be difficult to apply the results from this research in visual analytics. Another problem might arise from the fact that visual analytics, by definition, deals with large amounts of data. Most bias mitigation strategies suggest that users should look at additional data or check additional hypotheses. In contrast, users of visual analytics systems often want to get rid of large amounts of data to be able to concentrate on what they perceive as the really relevant facts. Practitioners in such areas also operate under time constraints. This also makes it difficult to motivate them to consider large amounts of data or too many alternative hypotheses. In addition, bias mitigation research does not consider the importance of background knowledge or expertise which is essential for decision-making in domains like medicine and intelligence analysis. All this makes it difficult to apply cognitive bias research in visual analytics.

In this chapter, an alternative method of bias mitigation is suggested: providing context and activate background knowledge. I want to point out, however, that this method is not entirely straightforward. There is research suggesting that providing context does not always help. More research in that area is necessary to identify how such systems should be designed to be useful.

Acknowledgements The research reported in this paper has received funding from the European Union 7th Framework Programme FP7/2007–2013, through the VALCRI project under grant agreement no. FP7-IP-608142, awarded to B. L. William Wong, Middlesex University London, and Partners.

References

1. Adderley R, Badii A, Wu C (2008) The automatic identification and prioritisation of criminal networks from police crime data. In: Ortiz-Arroyo D et al (eds) EuroISI. LNCS 5376, Springer, Heidelberg, pp. 5–14
2. Dimara E, Bezerianos A, Dragicevic P (2017) Narratives in Crowdsourced evaluation of visualization: a double-edged sword? In: Proceedings of the ACM conference on human factors in computing systems (CHI), pp 5475–5484 (2017)
3. Doppler-Haider J, Seidler P, Pohl M, Kodagoda N, Adderley R, Wong BLW (2017) How analysts think: sense-making strategies in the analysis of temporal evolution and criminal network structures and activities. In Proceedings of the human factors and ergonomics society 61st annual meeting
4. Evans JStBT (2007) Hypothetical thinking. Dual processes in reasoning and judgement. Psychology Press, Hove and New York, USA
5. Eysenck MW, Keane MT (1990) Cognitive Psychology. Lawrence Erlbaum, Hove and London, Hillsdale
6. Fiedler K, von Sydow M (2015) Heuristics and biases: Beyond Tversky and Kahneman's (1974) judgment under uncertainty. In: Eysenck MW, Groome D (eds) Cognitive psychology: revisiting the classic studies, Chap. 12. Sage Publications, pp 146–161
7. Gigerenzer G (2000) Adaptive thinking. Rationality in the real world. Oxford University Press, Oxford, New York
8. Johnson-Laird P (2008) How we reason. Oxford University Press, Oxford, England
9. Kahneman D (2012) Thinking fast and slow. Penguin Books, London, England
10. Klein G, Moon B, Hoffman RR (2006) Making sense of sensemaking 1: alternative perspectives. IEEE Intell Syst 21:70–73
11. Klein G, Moon B, Hoffman RR (2006) Making sense of sensemaking 2: a macrocognitive model. IEEE Intell Syst 21:88–92

12. Kretz DR, Simpson B. J, Graham CJ (2012) A game-based experimental protocol for identifying and overcoming judgment biases in forensic decision analysis. In: 2012 IEEE conference on technologies for homeland security (HST), pp 439–444

13. Kretz DR, Granderson CW (2016) A cognitive forensic framework to study and mitigate human observer bias. In: 2016 IEEE symposium on technologies for homeland security (HST), pp 1–5

14. Norman G (2014) The bias in researching cognitive bias. Adv Health Sci Educ 2014(19):291–295

15. Nussbaumer A, Verbert K, Hillemann E-C, Bedek M, Albert D (2016) A framework for cognitive bias detection and feedback in a visual analytics environment. In: 2016 Proceedings of the European intelligence and security informatics conference, pp 148–151

16. Sacha D, Senaratne H, Kwon BC, Ellis G, Keim D (2016) The role of uncertainty, awareness, and trust. in visual analytics. IEEE Trans Vis Comput Graph 22(1):240–249

17. Woll S (2012) Everyday thinking. memory, reasoning, and judgment in the real world. Psychology Press, New York, London

Printed in the United States
By Bookmasters